A CAT NAMED DARWIN

A
CAT
NAMED
DARWIN

*Embracing the Bond
Between Man and Pet*

WILLIAM JORDAN

A MARINER BOOK

HOUGHTON MIFFLIN COMPANY

BOSTON   NEW YORK

FIRST MARINER BOOKS EDITION 2003

Copyright © 2002 by William Jordan

Visit our Web site: www.houghtonmifflinbooks.com.

Library of Congress Cataloging-in-Publication Data
is available.
ISBN 0-395-98642-7
ISBN 0-618-38228-3 (pbk.)

Book design by Anne Chalmers
Typefaces: Minion, Scala Sans

Printed in the United States of America

MP 10 9 8 7 6 5 4 3 2 1

TO HOOVER

who led me the rest of the way

ACKNOWLEDGMENTS

With any book that takes as long to think and write as this one, the people who support you, who hold your hand and help you up and declare your genius, et cetera, et cetera—these special, essential people tend to recede into the fabric of daily, pedestrian existence and are sometimes forgotten when gratitude is officially passed out. Worse yet, the ones who assisted with the heavy lifting, reading the various drafts and suffering the endless drone of your voice as you read to them—they tend to become so familiar as to be taken for granted, like the ground or the air, and sometimes get overlooked. The fear of such inexcusable gaffes is my dark little paranoia. In the event I commit them, I am repenting in advance and vowing eternal chagrin. However, life goes on. Do we must, and do we should with sincerity and in earnest. Let chance have the rest.

In that spirit I offer my deepest gratitude to Michael Daves, Paul Ciotti, Patrick Pfister, Michael Parrish, Robyn Shirley, Bridget McCarthy, Jeannine Oppewall, and Sherry Virbila for reading and/or listening. Much gratitude and many thanks also to Doug Mader for his patience in explaining Darwin's medical history; to Daryl Mabley and Christine Belezza for offering other points of medical view; and to Douglas Domingo-Foraste for his kindness in correcting my Latin.

To Harry Foster, Elizabeth Kluckhohn, Peg Anderson, and

Sandra Dijkstra a grace note of thanks for the advice and support which, even though professional, still creates a sense of friendship above and beyond the job. And to Babette Sparr a double mordent of gratitude and affection for the friendship and advice that continues still.

And to my parents, the profoundest acknowledgment of their love and encouragement.

Contents

Introduction 1

1 Picking a Human Up 11

2 A Dog's Meow 21

3 Breaking Up 32

4 Inventory in England 46

5 Nuptials 61

6 Honeymoon Prognosis 64

7 Hope, Intimacy, Jealousy 71

8 Friendship and Equality 81

9 Hospice Care 95

10 Night Walk 112

11 Home Invasion 123

12 Sweet Epiphanies 150

13 Tender Mercies 166

14 *Missa Felina* 174

Epilogue 189

I was born a *Homo sapiens*.
Then I became a biologist.
Then, I became a cat.

You have no idea.
Read on, friend.

A Cat Named Darwin

INTRODUCTION

IT'S THE SOLITARY ONES who are most vulnerable — those of us who live by ourselves and have time, probably too much time, to think. It happens gradually, imperceptibly, like temperature rising or water seeping, and one day you find yourself noticing new lines, say, in his facial markings. You notice the way he greets you, nuzzling your outstretched finger, then sliding his mouth along your fingertip to the corner of his jaw. You notice the whites of his eyes as he watches you continuously, not out of wariness, but out of a gentle, calm trust we humans would call love. You notice the nuance in the way he moves, the subtle pauses and postures that express his own personality and distinguish him from other cats — and you hear the particular timbre of his voice and know intuitively with a crawling of the nape when he's threatened by another cat out in the wilds beyond the door. You realize at some point that his movements and gestures are a language, his tail wrapping gently around your leg, or his head pressing deliberately into your hand, or his mouth opening in a wide fang-bearing yawn of greeting as you walk into the room. The way he stretches forward and claws the rug, the little crook in the end of his tail, the unique tufting of his belly fur . . .

These quiet, introspective revelations are the gift of the cat to the solitary person, for the cat is a creature with whom you share solitude. A human being, on the other hand, is a creature with whom solitude is generally a failed relationship. With one the essence of success is communion. With the other it is communication. One depends on spoken language and rational intellect, the other on the language of gesture and intuition, and whereas communion with an animal is considered inferior to communication with a human being, the truth is, the need for companionship of any sort is a human species trait, and in the absence of a human companion, the mind grows like a vine around any living thing. The first time your mind grows around a cat, you do not realize you have fallen in love.

Communion with a cat takes time to mature, and it is irreversible. Those who find it are forever altered and cannot go back to the way they once were because the mind, the soul, the eye of self, arises from the physical substance of the brain, and that substance has been altered. The brain records experience continually in a running record, which is crucial to the working of conscious awareness. When you notice a new pattern on your cat's face — the stripes have always been there, but for some reason one of them now stands forth — this revelation occurs because the mind compares the current perception with visual memories. The longer you live with a cat, or any living thing for that matter, the more detail you see because the brain has had more time to record. This in turn sharpens the perception of detail in the present, the mind comparing present with memory and memory with present, back and forth, forth and back, in a resonating fusion of memory and instant that we experience as conscious awareness.

And how does the brain record these memories? We know in a general way that it does so through physiological changes. Neurons make new connections with other neurons; neurons recruit other neurons, so when one becomes active, its activity stimulates its immediate neighbors to join in; eventually a pathway forms along which the impulses of memory and perception run; complex chemicals are probably also involved in storing memories, and who knows how many other operations of brain physiology? This means that a physical mechanism — a neuronal machine — is slowly, gradually assembled in the brain to service the relationship, and details accumulate in the mind as more neurons, more synaptic connections are dedicated to your companion. Those who work at home and live the single life can easily spend 80 to 90 percent of existence with their animal comrades, which means that a very large mechanism indeed must be constructed.

You don't realize how pervasive this mechanism has become until your companion is taken ill; then the world cracks and crumbles around you. Its suffering becomes your suffering. When it lies in pain and silence you immediately grow depressed. If it shows the slightest sign of recovery, the sun shines into your soul and your spirits soar euphoric. In other words, the health of your companion controls your moods as if your nerves were linked directly together. You are fully aware of this influence, you just cannot control it.

And when your companion dies, the pain is almost unbearable. The longer and the deeper you love him, the greater the price in grief. It's as if part of your self has been amputated without anesthetic, which it probably has — literally — because the machinery needed to generate the miraculous subtlety and nuance you experience with your loved one is, in one ineffable instant, rendered moot. It has no more reason for being.

Without purpose, without meaning, that part of the brain devoted to your friend will now be altered. The gray matter is needed for life and the brain has now to be recast around the emptiness where you and your companion once lived.

Meanwhile the memory mind continues to operate as if your friend still lived, projecting images in all the places he loved to be, and you see him everywhere, lying on the bed, sleeping on your desk, jumping over the wall and walking gracefully to greet you on your return home. The fact is, those we grow to love continue to live in the synapses and molecules of memory and as long as we exist, so they exist as part of the brain. That is what happens when anyone loves anyone, or anything. It doesn't matter to the neurons deep in the brain whether those whom you loved were human or animal. The mechanism is the same.

When we are young and heading out into life, we are going to marry, of course, get a good job, raise a family, live a long, peaceful life surrounded by loved ones. Of course we are. What is there even to discuss? Not to marry, not to have a family, not to paint one's life by the numbers — that is not an option and it is not to be countenanced. It has to be denied. We must dream high when we are young, navigate toward a star, putting off for many years the fact that happiness is a state of denial. In case we need motivation, society presents us with a symbol of failure: the spinster with her cats, the aging bachelor with his dog. Failure in life, loneliness. Deep inside we pull back in pity and relief, thanking God that such will not be our lot.

Life, however, has a way of hindering dreams. People get divorced. They die from accidents or early disease. They pursue pleasure for a few years, and the few years become many; time passes them by. They fail to find the right one. Some discover they prefer freedom to marriage. For any number of reasons life does not work out as we had known it would, and people find themselves without human intimacy.

A cat then appears in the yard and we notice it lurking around. Without the urgencies of family responsibility, the notion of putting out food fills the blankness beneath the conscious mind, and the cat soon turns up every evening at the appointed time. One thing leads to another, and before long the cat comes into the house. It rubs against your leg, meows for food, jumps onto your lap. A name comes to mind. And you are on the way to conversion. Cat, dog, parrot, potbellied pig, hamster, canary, et cetera, et cetera — for any number of reasons, people find themselves with animals in lieu of humans, and if you could read their deepest feelings and thoughts, you would find that many of them are much happier than you might imagine. There are many paths through life, and some continue past the picket fence and the cozy bungalow of conventional dreams.

However, the vast majority of people do take the normal path, settling down with husband or wife, begetting a family. The world runs according to their values, as it must. The machinery of civilization with its industries, farms, hospitals, universities, government, all depends on people who course through life in that vast river of humanity known as the mainstream, accepting without question the traditional way in which we humans view ourselves against the backdrop of planet, cosmos, eternity, infinity. That view, with its self-promotional exultation, is essentially a Human Chamber of Commerce: "What a piece of work is a man, How noble in spirit, how infinite in faculty . . . in apprehension how like a god." Or, "God said, Let us make man in our image/ . . . and let them have dominion . . . over every creeping thing/that creepeth upon the earth." And ever since Darwin, "The Pinnacle of Evolution."

There is no understanding Life in its larger, planetary sweep so long as one adheres to this anthropocentric point of view, and we shall come back to this fact. Suffice it to say that the cat offers another way of seeing things.

⁓

All of which implies a set of core values essential to mainstream philosophy. These values are compressed into one hard, tough little three-word pellet of an expression: "Get a life."

"Get a life" most often implies that one is wasting time in trivial pursuits and ought to do something more significant with one's time. Keeping in mind that an extremist is anyone whose opinions are extremely different from your own, the mainstream person senses intuitively that those who cross the divide between animal and man have values that pose some sort of threat. In fact, the love of other creatures could, theoretically, revolutionize the nature of civilization. Civilization is manufactured in large part from living things, and if a majority of humans were to embrace all forms of life, treating them as kin with respect and reverence, the cost would come back to us in countless proscriptions and deprivations. Animal experimentation, animal husbandry, amusement parks, aquaria, and circuses would be strictly curtailed or eliminated altogether; the trade in ivory and ornamental furs would be eliminated; and 2 billion Asian men, deprived of tiger penis and rhinoceros horn, would be reduced to bleating castrati.

"Get a life" speaks to all of that. As a rebuke, it ranges in strength from gentle, patronizing reproach to utter, baleful hatred, depending on how radically the person addressed appears to differ from mainstream society, and when the lover of animals advocates animal rights, "get a life" becomes "fringe zealot."

The point being that it is natural and normal and inevitable for people sweeping past in the mainstream to belittle the lover of animals. Normal, mainstream people are not capable of understanding the mindset that lovers of animals evolve toward their companions for the simple, physiological reason that the brains and the minds of normal people grow chiefly around their spouses and children and only secondarily around their pets. Humans require the overwhelming share of attention. Animals get emotional leftovers. Mainstream human values, therefore, function as a social mechanism, like the invisible hand of Adam Smith, to glorify the human image of its species self. Those who take alternative paths must expect a certain level of prejudice and persecution and accept it, because that is how reality works.

Now if the deep love of one's animal companion is essentially a surrogate affair — a relationship that often grows in the absence of human companionship — and if society tends to look with raised brow and wrinkled nose at folks who go this road, that is not to say the rewards are necessarily inferior to those derived from the company of humans. In fact, one of the greatest of alternative rewards is the very absence of humanity. To live with animals is to recognize how obtrusive and harrowing the minds of other humans can be and to realize, ultimately, that innocence is nothing but the absence of the adult human mind. That is why animals are innocent, that is why infants are innocent, that is why sleeping adults appear as innocent as prior experience will allow you to perceive. By contrast, the companionship of a cat or dog or other creature requires no deceit and little conniving and allows us to indulge whatever fancy we will. Words cannot express what a pleasure this is.

Still, to have a creature at the center of one's world is the mark, according to mainstream standards, of a very little life, a life on the fringe. Ah, the irony of dwelling at this "fringe." You stand at the portal to another dimension, a universe so vast and rich and endlessly fascinating that once you have passed through, your perceptions of life, your values, your entire image of self, will be permanently altered. The cat sits upright and alert at the entrance to this portal, and you enter through its eyes, through those ecstatically clear, still eyes, passing into its mind, into its view of the world, into a comprehension of life that obliterates the human illusion and purges the Human Chamber.

The intimacy that humans crave at the center of love draws you inexorably into the animal's mind, yearning to *feel* how a different being knows the world. As time goes on, you begin to experience a sense of oneness, as if you actually *are* the creature you love, and when this occurs you have passed the point of no return. That which the animal gains, the human species loses, and your allegiance to *Homo sapiens* has been divided.

You have also been liberated. Now, for the first time, you stand at an emotional and intellectual distance from the values of humanity looking back at your own kind, and now you see *Homo sapiens* through the values of another species. How utterly self-absorbed we humans are, so narrow in vision, so parochial in interests, so driven by appetite, the infant mewling at the center of its own cosmos. Yes, and how *unsapient* our society appears from beyond the self, spinning faster and faster in a tarantella of quotidian chores, errands, duties, rushing forward in a fog of sightless schedules and commitments, and always, always poking, probing, questing for yet more efficiency in our appetite of appetites.

So it was, during my forty-fifth year on this glowing blue Earth, that a cat entered my house and stole my heart. When he beckoned me with a blink and a yawn, I followed him away on a journey to exotic lands and strange cultures. Why not? I thought. I had nothing to lose. The time was right. I had no wife and family to set my agenda and I could travel light, exploring places where those with children and the essential allegiance to *Homo sapiens* were not able to follow. And off I went, taking nothing with me but the spirit of science and the love of this little creature, because the spirit and the love were all I needed for the journey on which I had naively embarked.

Not long after we left, other cats entered my house, in particular Hoover and Little Grey, and as the bonds between us strengthened and our love and respect deepened, I became fluent in their language, and gradually it dawned on me that my companions had ulterior motives. They were not mere cats; they were philosopher cats. They were priests. And they had the agenda one would expect of philosopher priests.

"Come with us," they meowed in a chorus of sweet dissonance. "Humanity is a state of denial. Come with us and see thy species self."

"How dare you," said I with the righteous indignation of my species. "The human being is the pinnacle of evolution. Above the human there is nothing but the universe."

The cats did not dignify my reply with a direct answer, no doubt smiling inwardly with the sly recognition that the universe — God? — overarched every thing on the planet. They simply stared at me as cats stare. Then they gathered around and rubbed against my legs in the warm, soft friction of feline love, wrapping their tails around my calves and trailing them away with lingering affection as they turned and headed off.

For ten years we have traveled together, I following with the eyes of Gulliver, beholding at each turn the wonders of nature and the wonders of human nature, and these sights have changed me forever. What I once saw as the mainstream of human affairs, I now see as a navel fixation, arrant parochialism that obscures our true place in the body of a living, multispecific planet.

Ten years marks a natural cycle, however, and the time has come for me to tell the tale of where I have gone and what I have seen. *A Cat Named Darwin* is best regarded as a sort of travel writing, the collected letters home of a philosophic nomad.

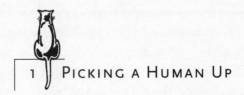

1 PICKING A HUMAN UP

THE FIRST TIME Darwin spoke to me I didn't understand a thing he said. I did, however, understand everything he meant. That is because he spoke the old language, the *lingua vertebrata* of posture and pose and cries without consonants that our animal kin speak from birth, and even though we humans have neglected this language in our tortured exodus to civilization, we still retain an innate ability to comprehend if we simply watch and listen, and feel.

I had gone out to empty the trash and was walking between the house and the old, rotting fence when I saw a big, orange, bull's-eye tabby lying in a bed of leaves beneath the bougainvillea bush just across the property line. He had been nesting there for about a week and usually ran when he saw me. This time, however, he held his ground and lay there, head resting on forepaws, staring into my eyes with a sullen, defiant glare that passed through my glasses, bored into my hazel-green retinas, and passed through the tiny black hole by which the universe enters the human mind.

I stood frozen, staring back, staring into those still, clear, metallic orange disks, into those black slits through time, into the ancestry of all who came before us.

Even though he was dirty and haggard, he was still a handsome cat with the classic bull's-eye marking on each side, a large, dark blotch in a light field, circled by a thick ring, and a white bib extending from his chin down his breast and over his tummy. I had seen him in the neighborhood many times and had taken little notice, but this time I could not take my eyes away, and I stood there, eyes locked with his. Then, driven by some primal urge I will never understand, I opened my mouth and meowed.

Immediately the cat raised his head, intensifying his stare, and meowed back. Then he stood, stretched, walked deliberately toward me, and squeezed through a hole in the rotted planks. As if obeying some extrasensory cue, I dropped to hands and knees so my face was no more than a foot above his head, and waited. He looked up into my eyes for several seconds, then slowly, carefully, raised his right forepaw and oh so gently touched my nose. Looking into his eyes, now a foot away from my own, I lowered my head still farther and watched with crossed eyes as the cat raised his face and touched his nose to mine.

I reached out impulsively to stroke his head. He leaned into my hand, savoring my touch as only the cat can. He rubbed against my thigh. I ran my hand down his back, and he arched into the stroke. Again I ran my hand along his back, and again. Then he turned deliberately around and, with the most nonchalant grace, bit my hand.

He bit my hand! It was not a savage, all-out bite, but it hurt, and I lurched up and back, tripped over a pile of newspapers, and fell clumsily on my back. The cat, apparently mistaking this maneuver for some sort of martial art, emitted a cloud of hiss and sailed over the fence in a single leap, tail lashing the innocent air.

My first reaction was to consider lethal force. No animal did that to me and got away with it. I had spent my early years on a farm and people on farms do not balk at taking animal life. My second reaction was a feeling of weariness. I was slowing down for middle age and something in me seemed to have changed. For the first time, vengeance seemed stale. It proved nothing but the obvious fact that we humans reign supreme. So instead of getting a club, I found myself extending my hand and cajoling.

"Come on, it's all right, meow, no one is going to hurt you, meow." I was sure the meow had no meaning, but I didn't know what else to say.

In a few moments the cat seemed to relax, and finally transcending his apprehensions, he squeezed through the hole in the fence and walked toward me, tensely suspicious.

At this point I must have entered some sort of trance, for I vaguely recall walking to the corner store and buying a can of cat food, walking back to my flat with the cat following close behind, climbing the stairs, opening the door, watching the cat enter and cautiously scout the room, watching his trepidations vanish with the aroma of food as he sat up like a bear, crying for service. Although I would have denied it at the time, I realize now I knew then that I had just committed myself to another living thing.

Ah, the blessings of ignorance. Had I known what the proper care of a cat entailed, I would certainly have walked away. But I didn't know and so now began the practical task of starting out. Relationships are always practical at heart and have little to do with romantic beginnings. It is a first-things-first, one-step-at-a-time, cross-the-next-bridge-when-reached process.

I took my first step with what might be called a fresh eye, since I had never lived with a cat and knew little of feline habits, but in fact that fresh eye peered out from all the mainstream values and attitudes of the late twentieth century. I was an animal liker, not a lover. As a creature of American civilization I had no idea what love and respect for other creatures meant, how it felt, what it required. As a citizen of the West, I assumed that an animal, no matter how enjoyable its company, was ultimately a commodity and not worthy of the priceless value we humans place on our own lives. In great part this is the legacy of Genesis, first chapter — "And God said . . . let [man] have dominion over . . . all the earth, and over every creeping thing." It is a view that culminates in that strangest of all environments, the holy ecology of Heaven, which has only one species, the human being.

My Western values were augmented, but also tempered, by the values of science. I had been educated as a biologist, had spent thirteen years at the graduate and undergraduate levels, and I had learned my lessons well, Ph.D. well. The first step in the scientific process is to observe what is there and *only* what is there. There was this cat.

The larger task of science is to see reality without the distortions of religion, culture, political ideology, and personal agenda. As a result, when I chose to look at life with a biological eye, I saw the laws and principles by which evolution has designed and crafted the living. In watching animals behave, I saw the strategies and calculations by which the living survive. In gazing over the natural landscape, I saw the objects and forces with which life must cope in order to succeed.

That is all I saw from the biological mindset. Sentiments like love and affection have no place in the workings of the scientific mind — unless they are viewed as mechanisms from an intellectual distance — and the bare fact was, this big, orange, dirty, hungry cat stood in my house waiting for me to feed him.

And what did he bring to the boarding gate? He brought a certain age. He had arrived in the neighborhood a year before, clearly a mature cat, and I thought at the time that he belonged to my neighbors across the street. He had walked deliberately toward me with all the nonchalance in the world, wearing a flea collar. I had reached down to pet him, but he merely tolerated a few strokes, then turned and walked away less than impressed, as magisterial as he had come.

I saw him periodically after that, as one would expect of a neighbor's cat. Gradually, however, he began to grow thinner, then gaunt. This seemed strange, since the neighbors took good care of their other cats, which were sleek and well groomed. One day I happened to meet these neighbors and asked them if the big tabby was theirs. They said no. They didn't know whose cat he was. He had probably been abandoned — people frequently leave cats behind when their lives change and they move away — and my neighborhood seemed to attract more than its share of these unfortunate strays.

And so what had been a big, sleek, handsome, neutered tom had slowly come to be this thin, gaunt creature of the streets, forced to pilfer food from the dishes of kept cats and dogs, to scrounge the alleys for scraps of refuse, and to fight for shelter and territory among the other cats without homes, driven, finally, to beg for food. Now he stood before me in what might be his last chance to find a decent life.

At forty-four, I too was coming to recognize my own mortality and for the first time feeling the isolation of the single life. I had come to a point where the self was not enough. The single life seemed hollow and listless and lay before me in a flat, overcast plain of existence. I yearned to escape it. I *wanted* to give up personal freedom and commit myself to another person, a prospect I had rejected all my adult life.

So there we stood, two confirmed bachelors, one facing the desperation of bare survival; the other, bare loneliness. I opened the can of food, dished half of the smelly contents into a bowl, and watched as the cat attacked it, redefining the expression "wolfing" it down. A feeling of pleasure came over me: vicarious gluttony. He quickly finished and meowed for more. I fed him more. Again he finished and meowed and again I dished out more, soon emptying the can. Then he went to the door and wanted out. I let him out. That evening, as darkness came, he wanted in. I let him in. He wanted more food. He received more food.

When he finished his second feast, he walked over to where I sat reading the paper and jumped onto the couch next to me. After staring steadily into my eyes for an eternal fifteen seconds or so, he turned about in several slow, tight circles, testing the firmness of the cushions with his forepaws, then flopped heavily against my thigh. There he lay for the rest of the evening, dirty and populated with fleas, and while he purred, I reveled in the feel of his warm, small, dirty body pressed against my flesh.

Finally, however, it came time for bed, which meant it was time for him to go outside. He was still just a cat, and as pleasant as our time together had been, cats belonged outdoors. *In rules and laws,* crooned my interior voice, *is Civilization. In this iron fist in this velvet glove lie those rules and laws.*

Heraus mit dir, said the beloved memory of my German grandmother.

I stood to my full height, expanded my chest with a deep breath, and pointed toward the door with a military stiffening of my right arm. The cat stared at me for several moments, then, as if he had heard my interior monologue and understood perfectly, he hung his head and walked dejectedly out, emitting a tiny, thin, pathetic meow. No human ever expressed resignation and despair with more pathos than this cat. I was still, however, an unregenerate member of modern society, and I tried to shrug off the waves of pity as a mere projection of human emotion. The door closed behind the poor creature, the latch clicked, and I had just committed the first blunder of many in our relationship.

I had underestimated the cat.

A loud meow then arose from just beyond the door. Another loud meow. I did nothing. A louder meow. More pity welled up, but it hadn't a chance of forcing me to reconsider my policies. Louder still. How much volume does he have? I wondered, but refused to open the door.

The cat responded with a relentless series of meows that went on and on, blending gradually into one long ululation that penetrated ceilings, walls, and floors and seeped into the rooms. The siege continued for at least half an hour; then, as suddenly as it started, it stopped. Nothing was ever stiller or deeper than the silence that followed. What was going on? Was he merely taking time to breathe? That question was answered with a strange scraping sound, as if someone was rubbing a piece of sandpaper against the surface. Again the sound came. And again. Then the pace picked up and I realized that the cat was pawing at the door, perhaps clawing. This went on for minutes before it stopped. No sooner had it stopped than the loud, caustic meowing started up. All through the night it continued, periods of billowing wails washing over my walls followed by bouts of small paws pummeling the door until finally, needing to sleep, I resorted to earplugs, which dampened the sound but did not eliminate it.

When I awoke the next morning the siege seemed to have ended. Light streamed through the windows and cast the shadows of leaves and branches against the walls, where they slipped silently this way and that across the whiteness. It was as if a spirit had departed. I opened the door a crack and peeked out. No cat. I opened the door farther. Still no cat. I opened it all the way and stepped onto the threshold, and just as I did, the cat slipped through my legs so quickly that I couldn't focus my eyes. I stood there with what must have been a lobotomized look on my face as it slowly occurred to me that I had been set up, the cat pressed like a commando against the wall next to the door, and when my guard relaxed he made his move with such perfect timing that it could not be blocked.

Into the kitchen strode the big orange cat, exuding confidence, expecting — knowing — it was time for breakfast.

Not that this changed the rules; cats still belonged outdoors, and one never budged on basic principles. Steel fist, velvet glove. That night the siege resumed, if anything with more determination on both sides. I inserted my earplugs and went to bed. The next morning the same vacant silence. This time I knew what to expect, but as I slowly extended my head to check behind the door, I saw a yellow Post-it, obviously from my neighbor across the landing. The cat's wailing, of course, would have been as audible to her as it was to me.

> Bill —
> I think there's a brain-damaged cat in the neighborhood.
> It yowled all night in front of your door for the second
> night in a row. Finally I threw a shoe at it.
> Diane

The cat was nowhere to be seen, and suddenly I felt a twinge of anxiety. Had he been driven away forever? A small chill of loneliness. He had spent so much energy in his campaign with such unwavering focus that he must be . . . and it occurred to me how desperate this little creature must be for the companionship of a human being, with its shelter from the real world. Then a strange feeling welled up in me and suddenly I wanted to call him, invite him in. But I had no name to call, so I simply whistled — a thin, quavering note from behind my teeth. About fifteen seconds later the cat appeared at the foot of the stairs.

Some context is in order here, because I grew up in a family of dog advocates who disliked cats and calculated their value against the gold standard of canine bonding and canine utility. We didn't dwell on the issue of cat versus dog, but if you added up the details over the years, the list would be downright damning. For example, with respect to that peculiar emotional sub-servience known as affection, which distinguishes dogs, cats seemed little better than reptiles. As domestic servants they were useless. At best they helped in rodent control, because they were unrepentant killers. They scratched furniture and urinated on rugs and wailed in the night. They neither guarded the house nor protected you from violent crime. They would not retrieve game or herd sheep or lead the blind. You couldn't train them — you couldn't dominate them and force them to your will — and therefore they were stupid. They were completely self-centered and did nothing for anyone but themselves; they were takers, not givers, thus they stood for bad values. Aside from the killing of rodents, their only benefit to man was in scientific experimentation. How anyone could bond with a cat was beyond comprehension; those who liked cats, let alone loved them, were probably limited in their emotional capacities.

So there I stood, unrepentant dogist and member in full standing of the Canine Nation, looking down without malice, experiencing the first tingling of a feeling I quickly suppressed, for I had no intention of assuming a long-term relationship with such a creature. In fact, I had no intentions of any sort. I was simply proceeding from moment to moment at the beck and call of impulses I had never before obeyed.

2 | A Dog's Meow

THE CAT WALKED OVER and rubbed against my leg, meowing to be fed, and it struck me just how thin, gaunt, and dirty he had become since we met a year earlier. His fur had lost its sheen and was matted on his back with crankcase oil. My German ancestry, however, ran deeper than my family values, and worthless though this little creature was, his unhygienic plight triggered a cleaning response. He would have to be bathed.

Practicality then reared its flat, scaly head. How to bathe a cat? Having had no experience in dealing with angry teeth and claws, I decided to ask the advice of friends who loved cats and owned many. Robyn, who lived around the block with three cats, referred me to a veterinarian who specialized in cats and ran a bathing and grooming service.

But how to get a cat from here to there? I did not own a transport cage, and purchasing one was out of the question for what was going to be a short-term relationship.

I called the vet, and the thin shaky voice of an old woman answered the phone, advising me to bring the cat in an old pillowcase. However, I soon discovered that evolution had designed cats to resist transportation in sacks. The cat and I negotiated the matter with some passion, but I cannot remember precisely how I convinced him to agree. All that remains are vague fragments of memory with images of claws hooking in cloth and wails of anger and desperation, of a cat held out at arm's length by the tip of its tail, where it cannot get you, of a cat rolled up in a towel. I recall a strong urinary odor as I drove. Later that afternoon, as I drove the cat home, the fur on his stomach and throat gleaming white, his markings a deep rich orange, a different odor began to waft from the sack, and my subsequent memories are very clear of washing the cat's rear quarters to remove the soil he had produced in sheer terror. Clearly, this creature had a deep-seated fear of veterinarians and automotive transportation.

The cat quickly recovered from the trauma, and that night, with a full belly, he curled up at my feet while I read the paper and watched TV. His warm, clean fur felt so comforting against my ankles. Later, when I put him out, he didn't cry and he didn't pummel and scratch the door. As I look back it is patently obvious that my life had taken a fundamental turn, and I hadn't a clue. The cat had abandoned his crusade for reasons known only to him, comprehending somehow that he had breached the walls to my soul, knowing in the reptilian roots of his brain that he had passed his trial by fire and found a home. The last person to understand this was me, of course, because as a human being I had the capacity — the glorious, essential capacity — to deny, without which life as we know it would cease to exist.

Over the next several weeks a pattern of existence began to emerge. Every morning I would open the door and find the cat sitting there, awaiting his food. I would feed him on the landing just outside my front door and, after eating, the cat would spend the day patrolling his territory and enjoying the rights of ownership, primarily sleeping in the sun or the shade and absorbing bliss. In the evening I would call him to dinner and feed him in the kitchen, after which he would walk into the living room and curl up at my feet or jump onto the couch and sleep next to me while I read or watched TV. When I went to bed, he went outside to enjoy the night.

I began to notice details of his appearance and behavior. His facial markings, for instance, led the eye on endless excursions through a labyrinth of fine markings. A line ran back from the outside corner of his eye and met another line running up from below to trace the outline of a mask. Five lines proceeded back from his forehead and converged in a cap of orange. They emerged from the cap and continued down to the base of his neck, where they coalesced into a single wide band that extended to the base of his tail. The tail, too, had its visual fascination, not for subtle complexity in its markings, but for the regular, half-inch spacing between the eight orange rings. But always my gaze returned to those thick circles of dark orange on each side that led the eye around and around, into a hypnotic trance.

I could not help but notice that the cat spent much time staring back at me, appearing to seek out my eyes or my face. I would walk away and sneak a backward glance and find him staring at me from behind. What this meant I had no idea, but the staring became a constant habit.

Then, of course, there were the fleas. They arrived in my flat like Ulysses' crew clinging to Cyclops' sheep, and the crew was impressive. Wherever the cat chose to sleep he left behind hundreds of tiny white eggs that seemed to glow against the black leather of the couch. There was no choice but to comb him as often as needed to remove these parasites — as much for my sake as for the cat's. I did not want to share my flat with vermin. I began to groom him every day and soon discovered that his reaction appeared to be hard-wired. In other words, he was incorrigible.

He tolerated, even appreciated, the combing of his head, neck, shoulders, and flanks, but any attempt to do his tail or hind legs provoked the most bloodcurdling threats of violence. This placed us in a dilemma. As a human being, I concluded that his legs MUST be combed. Fleas were having their way, and that could not be tolerated. Even though the cat objected, my superior overview of life trumped his right to dignity and he would have to endure a brief grooming each day.

So I called upon my superior human intellect to devise a scheme. I would wait until he was ravenous, and while he ate I would attempt to comb his hindquarters. This revealed that cats are able to yowl while frantically gulping food. They are also able to turn with blinding speed and rake their claws across the hand that feeds them.

Over the course of the next week I tried wearing leather gloves. The cat tried waiting before he ate until my hand came within range. I discovered that leather gloves were not the best protection against the cat's armament. Finally, having exhausted all options, I was forced to concede that there was no alternative but to call off my campaign against the fleas thriving in the dense cover of the hindquarters. And so the cat gave me a lesson in respect, revealing the fundamental truth that when push comes to shove, respect is a subcase of fear, that reprisal and respect cannot be separated.

Meanwhile, despite my cavalier presumption that cats meant nothing to me, this cat drew my attention compulsively. I could not be in the same room without glancing repeatedly at him, just as he gazed back at me, sometimes for hours. The thought never occurred that the more attention I devoted to his presence, the more memories my brain would store away.

One day, not more than two weeks after our meeting, I found myself thinking offhandedly about names. The cat needed a name. This had nothing to do with how long I intended to keep the creature or how highly I regarded him. My mind simply wanted something more specific to grasp than "that cat." To name, to name — yet another compulsion embedded in the human genome?

Now a name is a sacred thing. Just as a bad name is a curse that clings to one for life, a good name is a prayer, evoking the essence of the being, the creature, the *genius thingi*. A good name inspires its owner, often alluding to the traits of mythical figures, even gods, and if the name is truly sublime, it brings harmony and rhythm, poetry and music.

The cat's name came without conscious effort; a few moments later he was Darwin.

Dar-win. The word has an almost Celtic grace to it, with an apical *d* that bumps against the eardrum, then soothes the sensibilities with a gentle, rhotic *ar* that opens the soul to the wistful *win*, with its hints of gentle breeze and clean air. It also alludes to the gods, because those who study biology in any depth find themselves at the feet of a colossus upon whose work the conceptual integrity of biology rests, a man who has, with the passage of time, risen above the mortal world. "They look for the second coming," I once heard a biologist say. "They expect Him to come back on a cross. They blew it. He already came and went again. He called himself Darwin."

Perhaps. But such observations lie beyond my expertise. For me, the name paid affectionate homage to the Founder with the sort of gentle humor that Darwin the Charles would appreciate, coming back as a cat.

Having settled on a name, we began the task of getting comfortable with it. "Darwin" leapt easily from the tongue and I enjoyed saying it. Nonetheless, the name seemed a bit cumbersome at first, self-conscious, and it took several weeks for the image of an old, bald, bearded, thick-featured, white European male to merge into the image of a big, orange, white-bibbed, bull's-eye tabby. At the same time, Darwin the cat seemed to have learned his name; or at least he gave signs of recognition. I had merely to open my mouth and breathe "Darwin," and the sleeping cat would open his eyes, raise his head, and look at me.

I did, however, wonder about the cat's comprehension, considering that when I talked on the phone and mentioned his name, he gave not the slightest indication that he recognized or even heard his name. This may seem a quibble, but it has enormous implications in understanding the animal mind. In the theories of conscious awareness, one must be aware of one's self in order to recognize one's name; the self is widely thought to be the province of the human mind, as well as the mind of the great apes. Theoretically, cats, dogs, and all other creatures should not be able to "know" their names because they do not comprehend their selves.

In the case of Darwin, the evidence was inconclusive. He certainly seemed to know his name and reacted immediately when addressed. However, one night we watched a documentary on Darwin the Charles, and Cat Darwin lay there in slumbering bliss without so much as a flick of the ears while the narrator bandied his name about. "Darwin" meant nothing, apparently, when issued from the television. Did he respond mainly to the tone of my voice or to my inflections, or did he truly comprehend that his self had been tagged with a word, with a name?

On the other hand, what difference did it make in the quotidian course of life whether the cat responded to "Darwin" because he understood it was his name or because he recognized the peculiar timbre of my voice speaking a sound he had come to associate with good things?

As the days accumulated into a few weeks, the cat continued to surprise me by revealing a complicated personality. One of his more curious traits was to use his tail as a foot cushion. Whenever he sat he curled his tail forward and wrapped it around his forepaws. Then he carefully placed both paws on top of the tail, as if to insulate his feet from the floor.

Darwin's most distinguishing trait, however, was so odd that it took me several weeks to recognize. He usually expressed it as he waited impatiently at my feet while I prepared his meals. The ritual of preparation has a certain therapy in it, the clinking of spoon against bowl freeing the mind to wander while the faucet runs. In such a state of suspended intellect, one is unconsciously aware of neighborhood sounds. Somewhere children play and shriek and somewhere dogs bark and car alarms warble, and it was against this backdrop of domestic ennui that the barking of dogs began to rise on the horizon of my awareness. One voice in particular stood out with a gentle, sporadic bark that sounded oddly nearby. One day I happened to glance down just as Darwin opened his mouth and barked! I looked again to see if what I saw matched what I heard. Again his mouth opened and a small, high-pitched bark emerged.

How can a cat bark? This bore closer observation, and bending down I noticed that he was not actually barking, at least not in the way we humans expect a bark to occur. When we think of barking, we think in terms of human language and presume that a bark begins with a hard consonant like *b* or *p* — what the linguists call a "plosive" — in which the lips suddenly release the built-up pressure from the lungs in a pulse of sound.

But Darwin did not use his lips in sounding his bark. From what I could see, he formed the sound by suddenly compressing his ribs and (probably) his diaphragm, and this caused a burst of sound from the larynx. He merely opened his mouth to let the sound out and what emerged was a rather soft "Whu — , whu — , whu" — with a very short *u* that simply ended in mid-air. The plosive consonant, *b*, was not necessary because the sound began at maximum volume — a small shock wave — and that is what a bark is.

To verify this conclusion with a recognized master, I went downstairs to provoke the neighbor's dog — any pedestrian offended this creature — and observe the manufacture of a real bark. Sure enough, the dog used the same technique as the cat, opening its mouth just as the lungs and diaphragm forced an explosion of air through the larynx.

The implication was intriguing. The bark was not a bark at all; it was an *ark*. Actually, it wasn't even an *ark* because it had no terminal *k*. The *ar* sound simply ended abruptly in midair, which made the "bark" nothing more than a hard-headed, flat-rumped vowel. Another way of seeing it is to regard the bark as a compressed meow; the meow, on the other hand, is a drawn-out bark that starts low, rises gradually in volume, then tapers off to nothing.

Now if these notions and observations seem counterintuitive, it is probably because I was educated as a biologist, and biologists, like all scientists, assume that things are not as they appear. This is not taught to us as a conscious principle; rather we learn by example that if you stare long enough and hard enough at reality you will start to find all sorts of exceptions to what our culture has taught us to believe. The consequence is that science is fundamentally perverse; as a general rule, scientists find real pleasure in pointing out that traditional perceptions are wrong. There is much to be gained, because if people accept our claims, we gain guru power as keepers of the truth.

I had been well trained at the University of California during the late 1960s and early '70s, where I found refuge until the age of thirty, when I received a Ph.D. in entomology and my funding ran out. With no options left, I had to enter the real world and face the task of becoming a writer, which had been my dream since the age of thirteen (in part because I could work alone, at home, and not deal with the politics of a job). The process of literary metamorphosis took another fifteen years or so, because I had to throw my education away in order to woo the readership of good, mainstream people upon whom a writer's livelihood depends.

Or at least I had to throw much of it away — not all, however. Skepticism I decided to keep; perverse or not, it was the key to intelligence, and eventually, to wisdom, and I had grown to enjoy it. What better subject on which to focus my skeptical skills than a barking cat which revealed that dogs do *not* bark? I thought I'd let the mind out for a run, let it sniff around in our cultural illusions and flush out a few more contradictions.

It seemed that dogs and cats, and probably all our mammalian kin, did not generate consonants by using the tongue, lips, teeth. I could not think offhand of any that did — not mammals, not birds, not reptiles, not even our close kin the chimpanzees and the other great apes. They raged and sang at life using the open throat, and though they were able to produce abrupt sounds that seemed to start with consonants, these consonants were pretenders generated by the diaphragm, the larynx, and the lungs. Humans seem to be the only mammals that use lips, tongue, teeth, gums, glottis to produce consonants. All the others depend on the vowel alone.

These were restless thoughts and they arrived inevitably at the physical process of speaking, for the consonant has liberated speech from the bark, the meow, the howl, the ululation, and so on. If, for instance, we produced our speech as animals produce their calls, we would have to form each syllable with a separate pulse from the lungs. Speech would resemble a panting dog: *Speech* (lung pulse) *would* (l.p.) *re* (l.p.) *sem* (l.p.) *ble* (l.p.) *a* (l.p.) *pan* (l.p.) *ting* (l.p.) *dog* (l.p.).

The consonant revolutionizes all that. Instead of wheezing away in a breathless pant, we squeeze out long, resounding breaths to produce a continual flow of sound, much like a bagpipe, which we chop into sections with lips and tongue and teeth like some sort of verbal sausage machine, to produce syllables and words and phrases and sentences.

These observations ramified beyond language to the kinship of human and ape, the human and the chimpanzee sharing about 99 percent of their genes, which shines through in the family resemblance of arms, hands, digits, and general body form.

Then I thought of the chimpanzees I had seen, folding their lips back against the face, opening the mouth and howling, shrieking, hooting without the benefit of consonants, and I found myself seeing time from the opposite point of view and appreciating how far back we humans must have parted ways from our simian cousins. The anatomical machinery needed to produce this phonetic mastery must have taken evolution a long, long time, in particular the dexterity of the tongue and lips, the neural rewiring, and of course, the brain modifications needed to drive the anatomy of the mouth. The pharyngeal cavity, the sinuses, the epiglottis — a long list of alterations — and such alterations take eons, even epochs to accomplish.

I stared down as this big, orange, bull's-eye tabby sat at my feet, barking, and I thought idle thoughts. Provocative little creature . . . Already upsetting my Western view of life. Not what I expected . . . After all, this is a cat. What have I got myself into?

All that remained was to put this barking cat to some use. The next night I recorded his comments and spliced them into the outgoing message on my telephone answering machine: "This is Bill Jordan (pause). And this is Darwin" (the cat barks) "— my barking cat. Please leave calls and catcalls posterior to the beep (pause). Beep."

3 | BREAKING UP

OVER THE PAST thirty years or so, the younger generations have paired off in ever-growing numbers without the sanctity of marriage. To the young and callow it seems obvious that the ritual of yoking oneself to another human being in marital bliss is to accept arbitrary and antiquated values which lead not to marital, but to martial bliss. Why not simply live together and partake of the sexual and spiritual benefits without the embrace of responsibility? Sweet denial. The thought never occurs that maybe the burdens of commitment are the same, whether or not one has been formally bound by a priest, imam, Supreme Court justice, mayor, sea captain, or anyone else authorized to declare marraige.

I was enjoying the cat's presence more than I could admit without feeling burdened, and this pleasure made me aware of things that needed fixing. His teeth, for instance, were encrusted with tartar, and this got my conscience on edge: those teeth had to be cleaned. On the other hand, the job was probably going to cost more than I could afford and I had to keep my values straight. How much would it cost to get the cat's teeth cleaned?

32

Anything more than about twenty or thirty dollars was overly expensive, and if veterinarians were anything like other doctors, the cost could be considerably more. First, though, I would have to find a good vet, one whose office was nearby.

I called a friend who loved animals, worked at the local university, and, because of her formal education, devoted much of her life to lamenting the uneducated state of the popular culture. She recommended a vet whose office was only two blocks away. Two blocks were two blocks, however, and required that I insert Darwin into a pillowcase again. Maybe the time had come to invest in some sort of transport cage. I went to a pet store and found just the thing: a cardboard box with foldup handles and designed with cats in mind. Five dollars. Sold.

The good doctor turned out to be a tall, dark-haired man of about thirty-five and so exceedingly handsome as to seem at odds with the stolid, selfless, practical spirit one would expect in the givers of medical care, people who cannot worry about personal appearance when the job calls for rolling up sleeves and plunging the hands into open wounds, open bowels. I didn't dwell on the man's appearance, presuming that he must have proven his mettle during the course of a medical education, but I did laugh inwardly at his name, which happened to coincide with the name of an alcoholic drink.

Our conversation began as I explained to Dr. Grog — the name I shall use to avoid libel — that I wanted to get Darwin's teeth cleaned, that he was not exactly my pet, that he was just hanging out at my place, but that cleaning his teeth seemed like a good idea. Dr. Grog conducted a cursory exam as I spoke, prying open Darwin's mouth and peering in, then listening to his heart and lungs with a stethoscope. It struck me how docile the cat was, submitting to the indignities of examination as if he were half asleep. Wrinkles then appeared on Dr. Grog's brow, a grave look came over his face, and he said softly, seriously, "I'm afraid he has a murmur."

A chill crawled up my back but did not linger.

"So — what does that mean?" I asked, suspicious of what a heart murmur had to do with the cleaning of teeth. Murmurs are a common phenomenon in people, and many live long, long lives with them. Why would the case be different with cats?

"Well," said Doctor Grog, "I can't clean his teeth until I run some tests on his heart."

"What would that cost?"

"Around two hundred and fifty dollars."

"Why do you want to test his heart if we're just going to clean his teeth?"

"Well, because bacteria could get in through his gums, which can get cut during the cleaning process, and infect his heart valves. That's what the murmur means — something is wrong with his valves."

I knew from having studied biology that the chances of heart infection were small — after all, the cat was designed by nature to eat flesh and bone, and when you eat bone, your gums are likely to get cut from the broken ends. Certainly cats had evolved the ability to cope with such injuries and with the bacteria that came along, and I could feel my back arching at the veterinarian's patronizing tone, contrived, probably, to manipulate people who truly loved animals and threw themselves in fear and faith at the good vet's feet when their pets fell sick. Well, he had not bargained on a tough nut like me, a biologist who listened to all doctors with a skeptical ear.

"Look, there's no way I can afford two hundred and fifty dollars for heart tests — and that's on top of the cleaning fee. By the way, what *is* the price for the teeth?"

"Well," said Dr. Grog, getting huffy, "I'm not doing his teeth unless he has the proper tests. It's a matter of policy."

"For crying out loud, the cat is just a stray."

This goaded the doctor into revealing his true colors, and they spanned the entire spectrum of the color green. "You wouldn't say that if you came home some night and found Darwin having a seizure."

"You don't know me, friend," I snapped back, driven more by male bravado and spite than by true conviction. "If the cat has a seizure, it will just have to go ahead and have it."

No snake-oil vet was going to manipulate me with cheap sentiment. But as I lugged Darwin home in his cardboard container, the question kept nattering away inside my head. What *would* I do if I found him in the throes of cardiac seizure? Dr. Grog's tactics were clearly insidious, shamelessly insidious, but why should the question bother me if a stray cat were truly as worthless as I professed to myself?

I decided, after adding up my feelings and values, that Darwin's teeth would have to wait for another veterinarian — another trip to the hall of medicine. The experience, however, left a hollow feeling where a solid sense of righteousness ought to be. Darwin and I were back at square one in our quest for a doctor we could trust and respect, and I had no idea how to go about finding one.

⌒

We resumed the rhythms of existence, and Darwin prospered over the next few months, his frame filling out and his pelt growing thick and shiny. We grew more and more familiar in our daily routine, and life became a series of intimate episodes.

One of our more endearing interactions, for example, occurred whenever I took a shower, an event that Darwin found irresistibly intriguing. I would turn the water off, reach out for a towel, and find the big orange cat sitting on the bathmat, staring up at me. Apparently my ablutions showed through the shower curtain but were blurry, unfocused, and therefore novel. To focus for the strike was the *sine qua non* of cat existence, and blurry movement was probably impossible to ignore.

Familiarity, intimacy, the communion of ennui — we proceeded toward the inevitable. One evening after dinner, I looked down at Darwin and he looked up at me. We gazed at each other for several seconds. Then he raised his tail so it was sticking vertically into the air with a small crook at the end and took several small mincing steps around my feet. Without thinking, I reached down, scooped him up, and cradled him in my arms, nestling his head in the crook of my left arm, supporting his rump with my right arm.

The move could not have been performed more smoothly had it been rehearsed for months, and it occurred to me as I gazed down at this small creature that in fact the move *had* been rehearsed, countless times over countless generations. How many men, women, and children have swept how many ancestral cats into their arms and rocked back and forth in the embrace of the ages? I stood there absorbing the warmth from Darwin's body, savoring the feel of his fur on my skin, and understood then that what I felt was not confined to the present but extended back in time like a puppeteer's string, connecting our bodies and minds so that actions in the past traced themselves out here and now, on the cutting edge of the instant.

Unmoved by my philosophic reveries, Darwin closed his eyes and left soliloquies to me. I was happy to oblige and thought of Shakespeare. What a piece of work is a cat. How noble in spirit, how infinite in instinct. In form and movement how like — an angel? (I wasn't so sure about *that* — but not to quibble.) In voice how like a siren, which is to say how like an infant.

Yes indeed, this furry little face had its moments. Its cuteness spread through my substance in waves, and for a moment the sheer wonder of creation muted the noise of conscious thought. It was as if the ancestral creatures from which the mind is fabricated were communing directly with the creature in my arms, and I simply let the beauty of life flow through. The large, pure eyes, the finely chiseled nose, the velvet ears. What was it about this face that affected me so deeply? Without question it had to do with the illusion of kinship. As a general rule the more closely related two species are, the more similar they look and the more features they have in common. This holds for behavior as well as physical traits and has a most interesting consequence: communication — or, in the case of us humans, the yearning for communication. The features of the cat are similar in basic ways to the features of the human infant and that is why cats fit so neatly into the maternal-paternal reactions that evolution has fashioned into most of us.

Mammalian kinship was plucking my strings. I gazed down at Darwin and rocked him gently, his body fitting perfectly into my clutches, his small, furry countenance delighting me in the same way, I suspect, that a baby's face delights its parent, between diapers. Who needs an infant, I thought, not having children of my own, if you have a cat? The razor-sharp innocence of each slices into the soul with each cry in the night, with each mewling, whining request.

⁓

Then, in one brief moment, everything changed.

We had finished dinner and retired to the living room for the nightly pastime of reading the paper. Darwin was indulging an expansive mood and first lay down against my feet, then turned over onto his back and began to roll and paw the air. I took this as an invitation to join the celebration and rumple the fur on his belly. He writhed in ecstasy, squirming and twisting and rolling from side to side. Then, without warning, his body seemed to fold over my hand. Before I could even begin to pull away, his claws had locked onto my skin, his rear feet were cocked to rake my arm, and his mouth was fitted over my wrist, ready to bear down and drive fangs deep into me.

At the same instant an icy logic seemed to lock my mind, and I resisted the reflex to jerk back, forcing my hand and arm to go limp. There we paused for maybe two or three seconds, and then, as if realizing what he was about to do, Darwin loosened his grip, pushed my arm away, and leaped back in one motion.

I sat there with sweat oozing from my pores, the television flashing and yammering in the background. The speed and power with which a cat can move astonished me, and I sensed in the animal labyrinths of my brain how close I had come to severe lacerations, to the virtual slaughter of my arm. Then I noticed a brown smear of fecal fluid glistening on the underside of my forearm where the base of Darwin's tail had pressed against my skin as his body had wrapped around my hand. If I needed physical evidence of how close I had come to the edge, there it lay.

I stared at the creature who only seconds before had been reveling in affection, and it occurred to me that perhaps he was afflicted with some sort of involuntary reflex that could fire off without warning. The room seemed empty, bleak. Years before, an argument with a girlfriend had flashed into a brief physical altercation and I had experienced the same feeling of spiritual vacuum; now, just as it had then, came the awful feeling that things were over for us. This, however, was a notion to sleep upon, for the relationship was on trial.

The next morning I awoke well baked from a night of hot, fitful slumber and knew that the decision had been made. The cat would have to go. Things had been said that could not be taken back; injuries had occurred which would not heal. Furthermore, having no experience with feline relationships, I thought of my neighbor's note and it struck me that perhaps the cat was indeed damaged in the brain. It seemed ill-advised to ask for further proof.

I sat in bed, aching inside at the finality of it all, and Darwin meowed to come inside for breakfast. He entered as if nothing had happened, expecting with his customary impatience to be fed and giving no indication that he remembered the incident. I fed him, of course, although with some trepidations.

Then I began to consider the practical details of removing Darwin from my life and I realized I faced a quandary of sizable dimensions. The most obvious choice was to take him to the animal pound, where he could be adopted by someone else. However, even though I was not experienced in these matters, I knew from my cat-loving friends that the pound was for practical purposes a death sentence. I quickly rejected it. What has happened to me? I thought, as I looked down from above while Darwin ate. Why does the thought of parting ways bother me? I didn't know. I knew only that it did matter to me.

Very well, one of my friends would take him. They were more experienced in the ways of cats and would know how to cope with the peculiarities of his temperament.

I soon found out that my friends could not or would not take him away. They all had more cats than they should properly have. I began to call friends of friends and put out the word that I had a wonderful cat for someone to adopt. After a month of active campaigning, Darwin still had not attracted an offer, and finally the truth became clear: people did not want older cats; everyone wanted to start out fresh with a kitten.

Meanwhile metabolism went on. Mouths had to be fed. Sleep had to be had. Life continued to rotate through its ancient cycles. But Darwin and I were going through the motions of existence, or at least I was. Psychologically, emotionally, I had withdrawn into my self and refused for the most part to acknowledge the grace of Darwin's presence. When he did try to cuddle after dinner, I stroked him gingerly, tensed to leap back at any sign of aggression. He showed none, however. He seemed more docile, as if our close encounter had exorcised some sort of demon, but it made no difference.

Our relationship had arrived on some high desert with both of us trudging along parallel paths, looking straight ahead, each absorbed in his own reality, with nothing but the daily encounters for feeding and perfunctory contact scattered across the landscape like creosote bush and ocotillo. Darwin was completely unconcerned. So long as food and water came from the great source in my flat, his spirit depended on nothing but his whims.

⌐

So we led our separate lives, I following the contrapuntal score of human nature and Western culture, Darwin following the ancient script of feline genetics. The clear, cold air of this emotional desert, made for rational detachment, and as the days passed I began to notice patterns in Darwin's daily habits. I began also to see the differences between the behavior of a neutered tom and that of the feral, testicle-toting *hombres* that infested the neighborhood. Mainly it was a matter of degree, of intensity.

For instance, territoriality. Patrolling his fiefdom was the reason for Darwin's being. I had always assumed the loss of his pearls would liberate a male from the curse of testosterone, but I learned that in Darwin's case, this was nowhere close to the truth. When not patrolling, he was sleeping strategically, bedded down in places with a clear view of his boundaries or with downwind notice of approaching foes. Clearly, this behavior was wired into Darwin's brain and while high hormonal levels would intensify it, undoubtedly testosterone was not the fundamental cause.

These territorial tendencies brought to mind incidents I had observed over the years. Since my office windows overlooked three backyards, I could not help noticing the relentless campaigns of tracking, ambush, and skirmish to which the neighborhood toms dedicated their lives among the gardens, shrubbery, and sagging clotheslines. In particular I remembered one tough, old, battle-scarred veteran who had owned the land below my flat for about two years. He wore his lumps and scars and threadbare fur like full military dress. His ears were shredded, his brow balding with scars or mange. His gray coat always looked dirty, and he glowered at the world through milk-white eyes. Even I shuddered when he appeared from beneath some bush and skulked off in silence after some clean, fluffy, innocent house tabby out for a stroll in its owner's backyard.

One day a challenger arrived. He was a young, long-haired standard tabby, not nearly so large and probably no match for the old warrior in an out-and-out brawl. That didn't deter him. Despite his physical limitations, he stalked his older foe relentlessly, and I peered down from above for hours, as the two moved here and there, the challenger always following, gradually closing the distance between them until finally, when the old warrior was least expecting it, the challenger would jump him from behind. A large, writhing ball of fur would then roll about on the ground, emitting large wads of hair and horrible screams and air-rending hisses, and from time to time you would get a glimpse of what was going on inside. Usually the young one fought on his back, kicking furiously with his hind feet at the older cat's belly, but clearly getting the worst of it. Finally they would stop out of sheer exhaustion and glare at each other, crouched, tensed, waiting for the slightest move on the other's part, and disengage oh so slowly, as if their bodies were thawing. The old warrior would walk deliberately away, ears laid back to detect any sounds from behind, and turning now and then to glare back with the most baleful malevolence.

At some point I realized that the old tom had not appeared for several weeks, and I never saw him again. What became of him, driven from his home, growing older, weaker, slinking through a life of trespass on the land of the ever younger and stronger toms who tracked and attacked him without mercy? How did he die? — for almost certainly he had. Youth had been served. I shuddered to think of Darwin condemned to such an end and appreciated a bit more clearly the prospects he was facing when we met.

What emerged was a lesson in the power of simple, undiluted persistence. Even though outgunned and getting the worst of each encounter, the young tom licked his wounds and promptly set out after the old tom again, implying to his foe that unless he was physically disabled or killed, he would never stop. He persisted his way to the top, and I could see the demoralizing effect of this never-ending harassment over the course of hostilities. In fact, early in the skirmishes, I could *feel* the dominance passing from the old to the young. In this case dominance was not to the bigger and stronger physique, but to the tougher and more ruthless will. Youth was also a factor, and perhaps the young cat could sense that this old warrior was running on old batteries, unable to recharge and recover as fast as his young challenger, thus vulnerable to a long campaign.

Another lesson was the remarkable sense of geography by which the cat assesses the world. In stalking quarry, you need a mental map of the land to tell you where your quarry is likely to go, and this allows you to pick a place to lie in wait. Cats were masters of the ambush; I observed this mastery almost daily.

Darwin, too, displayed a fine sense of spatial setting. Once he batted a golf ball under the couch, where it disappeared. A dog encountering this dilemma for the first time usually tries to get directly at the object; it might reach under with a paw and, failing in that, course back and forth in front of the couch in growing frustration; finally, having rounded the corner in its ever-extending dashes, it discovers where the ball lies. On the next occasion, the dog remembers this route to success and proceeds immediately around the corner to get the ball, but its behavior gives little evidence of topographic comprehension.

Darwin, however, wasted no time in useless pacing. He sized up the situation in a long glance, stood up and, without hesitating, walked around the end of the couch to retrieve his errant toy. He knew exactly where he was going.

Another time, as I walked back from the post office and approached my flat, I saw Darwin sitting on the sidewalk in front of my landlord's house (behind which stood the garage and the two second-story flats). This was the first step in a ritual we had evolved in which he would wait as I approached and, when I was within about twenty-five feet, turn and scamper down the driveway to wait for me at the foot of the stairs.

There were two routes to my flat. Approaching from the north, I could walk past the landlord's house and down the driveway, which ran along the south property line, or I could turn off before the house and walk along the narrow passageway between the wall and the north line. The two routes were parallel.

The devil was in me that day, and out of sheer perverseness he decided that I should have some fun at Darwin's expense, justifying it in the spirit of science. I decided to see how thoroughly a cat could comprehend the layout of his land.

When I approached the house, I could see Darwin preparing to dash ahead as I turned down the driveway in the choreography of our daily ritual. Then I began the experiment. Instead of proceeding to the mouth of the driveway, I stopped at the north property line, turned right, and walked toward the passageway between the house and the rotting fence. Without hesitating, Darwin turned and dashed down the driveway in the same direction I was going toward the flat. Clearly he understood that his course ran parallel to my own on the other side of the house and he intended to meet me when I emerged from my alternate route.

Then I added another twist. Instead of continuing along the north passage, I stopped and traced my steps back to the street. Then I walked to the driveway, turned right, and proceeded back to the flat as I normally would have done. I passed the rear of the house, looked north, and there sat Darwin, his back toward me, waiting for me to arrive there, as my fake move had indicated I would.

Darwin got into relatively few fights, and when he did, they were not the savage brawls that true toms are fated to endure. He grew more and more mellow, spending much of his time slumbering away in various nooks he maintained here and there around the immediate backyards. Some were rather ingenious, and one in particular stood out. My landlord's son had drydocked an old Volkswagen in the space between the house and the garage, draping it with a Cadillac cover whose extra folds sometimes caught the breeze and billowed out. It took several months, but one day I realized that the fat, smooth bulge in the cloth below the front bumper was not billowing the way a billow ought to billow. On prodding it with the point of my shoe, I discovered that the billow was filled not with air, but with the prosperous body of whom we speak. Somehow, in his daily rounds, Darwin had discovered that the fold was also a hammock and it became his favorite bed for many months.

4 INVENTORY IN ENGLAND

SUMMER WAS NOW DRAWING to a close, and I received a ten-day assignment to cover a wildlife film symposium in England. Ten days is a relatively minor trip, and in the past I had simply gone off. Now, to my consternation, the matter of Darwin's welfare came to mind. And it came to mind again and again. I could not enter a private thought without Darwin creeping in a few minutes later, gazing up at me with a kink of expectation in his tail and that earnest, guileless calm in his stare. He now depended on me for food, and whether I liked it or not, I could not flick away the notion that I was obligated. As if to underscore the point, he began following me across the street to my friend Doug's flat, an act whose significance I could not ignore. By following me, Darwin was leaving his territory and entering enemy ground, and this meant he was depending on me for protection. I had, in Darwin's eyes, become his mommy. Not his daddy, because male cats do not participate in the rearing of their young, but his mommy.

The irony was not lost on me. I was acting more and more like an animal lover, although my values and perspective had been adjusted to reflect the changes. Besides, there was nothing wrong with caring for an animal, just as long as you didn't care too much, and even though I had not officially renounced my decision to find Darwin another home, I could not throw Darwin back to the streets, even for ten days, no matter how intimidating his tendency to bite and no matter how insignificant he was in the grand overview of human society.

The southern California weather now became a point of consolation. We could expect a monotonous queue of beautiful, sunny days, and that, coupled with the fact that Darwin was an experienced, street-savvy cat, helped to stifle my worries. He really ought to be all right for ten days. Which left me with the matter of finding someone to set out food and water twice a day. Even once would do, but the food dish *would* have to be set in a tray of water to prevent ants from taking over.

The neighbor across the hall was the logical choice, and had Diana still lived there the choice would have been ideal, even if she did think Darwin's brain was damaged. Diana was honest and responsible and led an orderly life. Every night she came home from work and every morning she left for work, and to place food and water outside the door would have fallen neatly into her routines.

She had, however, moved away shortly after Darwin co-opted my flat, and a young woman named Berdy moved in. Berdy was a strange, elfin little creature whose eyes appeared permanently dilated. She scurried to and from her flat with head down and face averted, and even though she responded with warmth and verve when hailed directly, she never initiated conversation and never spoke more than a few perfunctory words. In her late twenties or early thirties, Berdy appeared to hold a job of some sort, but she still depended for emotional support on her parents, who lived across the street, and it was clear to everyone in the neighborhood that she was not quite normal. This conclusion was strongly reinforced by the one glimpse I caught of her flat. Papers, trash, food wrappers were strewn everywhere, which left me wondering if she could manage an extra daily chore.

The fact was, I had little choice in the matter, for no one else in the neighborhood was available. When I asked Berdy if she'd be willing to feed and water Darwin, she agreed brightly. I did not sense the irony at the time, but her parents were retired missionaries, all three embraced Christianity with fundamental faith, and I had asked them to care for Darwin.

Arrangements made, I went off to England, expecting to forget the shuffle of domestic life, immerse myself in the art of wildlife filmmaking, and absorb some civilizing English culture. For the first few days I did just that, enjoying the wonders of modern wildlife films and the earnest company of wildlife filmmakers. In the evenings I reveled in the mythical ethos of Roman baths, the charm of cobbled streets, the rows of Georgian buildings crafted from stone, the music of the dialect with its round, rhotic *arrr*s. My life in California ceased to exist.

It ceased to exist, that is, until Darwin arrived. Most of the symposiasts were staying in the dormitories at the University of Bath, the school term not having begun, and we took meals in the cafeteria. On the third day Darwin appeared under the breakfast table, sitting between my feet. I could almost feel his plump little carcass pressing against my leg. It was time for his breakfast and there he was, gazing patiently up at me. Considering his skill at infiltrating my life, I should have known that he would stow away in my mind, and there was nothing, apparently, to be done about it. Furthermore, he had license to interrupt my life and he followed me about the symposium, showing up in front of me whenever he wished, often during lulls in conversation, sometimes in the middle of intense dialogue.

Then one night an incident occurred that removed any doubt how deeply Darwin had burrowed into my mind, into my soul. Probably the first thing that struck me on arriving in England was the precision and grace and casual certainty with which the English use English. They are masters of it. To watch and listen as English symposiasts arose one by one in a packed auditorium and spoke without notes or rehearsal for minutes on end was a wonder to behold; you could almost see the sentences proceed from their mouths fully parsed, clauses and phrases trailing from their proper antecedents — whole blocks of argument flowing by in stately paragraphs like the opening of a motion picture — and I found myself glancing around now and then, looking for the teleprompter: no one could speak that well.

This precision and grace reached its highest form in what could be called the surgical insult, a skill that appeared to be practiced as a form of recreation. Contestants would meet, often in some chance social encounter, and without warning one would calmly sever his opponent's aorta with a vicious insight to his personal deficiencies. Without blinking, the opponent would just as calmly insert a devastating comment into his opposite's liver, then remove it.

At first it was breathtaking, almost exhilarating, to observe these English sophisticates revel in their skills. How could people *think* of such brilliant, cunning repartee in the heat of battle, how did they find such discipline to refrain from offering knuckle sandwiches, as would be the response in American society. But very soon it became clear that you wanted to watch these displays from a safe distance because nothing was more desirable to these English masters of the verbo-martial arts than the wide-eyed American. And nothing was more terrifying to the American than realizing there was nothing between us and an English man, or woman, gripping in each fist a twelve-inch sarcasm with a serrated edge. We marched, we Americans, in cringing flocks to the slaughter at this wildlife film festival. The letting of blood took place on the third evening, in the form of a Hooley.

The word "Hooley," which had been coined at an earlier symposium in Ireland, is one of those onomatopoeic words whose rhythms and tones somehow express its sense. This Hooley was supposed to be a lighthearted celebration in which the participants cast modesty to the wind, aided and abetted by significant amounts of beer and wine, and performed whatever it was the performer felt shameless enough to do. Musical instruments, songs, poetic recitals, theatrical skits — all were welcome.

As it turned out, theatrical skits were the most popular choice, probably because the majority of the symposiasts were English and resorted reflexively to their cultural tradition. The skits were necessarily of a primitive, farcical nature, some of them impromptu, and were performed in a cavernous, dimly lit cafeteria where the amplified merriment echoed in screeches and booms from the bare walls and dark corners. It soon became clear, however, that this particular Hooley was not to be the innocent, lighthearted affair advertised, because the topic of main interest was American culture or, rather, what the English participants perceived as the lack of it. The method of celebration was going to be satire. Relentless, bellowing, nasty satire.

Now I am no knee-jerk defender of all things American. I shake my head in resignation at the growing vulgarity of our society and the deification of money. Money has become morality. Whatever makes money is right, and good, even spiritual. But whatever our cultural flaws, I discovered that night that the country of my citizenship is my identity. No matter how enlightened and decently different I fancy myself to be, in the eyes of other people I am still American. There is no choice in the matter. It is my country, right or wrong.

At first I was able to shrug off the snidery in the spirit of objectivity, even good-natured joshing, but gradually the atmosphere began to nettle the skin. Then it began to draw anger. Finally, having neither the confidence nor the skill to take the stage and rebut the aggressors in a civilized, sophisticated manner, I took my metaphoric dollies and stalked from the hall with hoots and yawps of cultural derision ringing in my ears, and retreated to my dormitory room, determined to sulk and feel sorry for myself.

"I feel like a grown son," I wrote in my journal, "whose English parents have grown senile in their old age and, despite the fact they need my economic support, do everything they can to insult me. I'm a guest here, for Christ's sake. We would never treat an English visitor like that" — overlooking the British tourists who strayed from the safe roads of Florida and ended up murdered for a few traveler's checks. I continued my literary rant for a short while, but writing, even when passionate, is hard work, and the steam soon stopped hissing from my ears.

I was sitting there, depressed, when Darwin appeared. He sat by my feet, looking up at me, oblivious to the world of human cunning and malice, and regarded me with the clear, pure honesty that only those without intellect can know. Here, in a moment of stress, I found myself thinking of a cat, and it struck me how enormously comforting the thought was. How was he doing back in Long Beach, sleeping in his bed of bougainvillea leaves? Was Berdy living up to her promise and feeding and watering him? Was he yowling and clawing at her door as he had at mine? Even these worries were comforting.

Then another apparition appeared; and before me stood an old gentleman with a massive, overhanging brow and a pate as bald as a light bulb. I was in the spiritual presence of the man whose name the big, orange bull's-eye tabby had assumed.

Charles Darwin is as close to a deity as the modern biologist is likely to get in this secular age, something akin to a saint but inclined toward birds and beetles and worms and coral reefs and fossilized forebears — a druid, perhaps, or, in this case, the Arch-Druid. England, as the nation of Darwin's birth, has a certain holiness. As I sat in the dormitories of the University of Bath (which had only showers), in a town where the spirit of Romans, Celts, Anglo-Saxons, Normans, and jolly friars arises in miasma from the earth, I paid homage to the man who had revolutionized Western thought, wondered what sort of man he was.

I had read his autobiography as a graduate student of biology and found it, at just eighty-eight pages (Oxford University Press edition, 1967), a haiku to brevity, simplicity, humility, moral character, and intellectual honesty, a masterpiece of compression. The lack of pretension had astonished me, immersed as I was in the world of academic competition, and certain passages had slipped into the grooves of my brain. Now they extracted themselves and came back to me as a sort of fragmented monologue directly from the Arch-Druid's mouth. He stood in profile, looking across the room without expression, and spoke to me in my own interior voice, the voice with which I had read his words the first time.

"I have attempted to write the following account as if I were a dead man in another world looking back at my own life. Nor have I found this difficult, for life is nearly over with me. I have taken no pains about my style of writing."

Darwin was sixty-seven when he started the autobiography, an age when confessions come naturally, for political reasons if you believe in an afterlife, and for liberation of the conscience if you don't, and he started at the beginning of his life, sweeping sins before him.

"Once as a very little boy, . . . I acted cruelly, for I beat a puppy I believe, simply from enjoying the sense of power . . ." That this incident affected him so many years later revealed the depth of Darwin's compassion, expressed here as guilt, because he was a man of such emotional sensitivity that poetry might have seemed his natural calling; his genius, however, lay in a rationality that equaled and probably exceeded his sensitivity. He spoke on in the flat tones of the rational intellect. " . . . but the beating could not have been severe, for the puppy did not howl."

A Victorian emphasis on character pervades the autobiography, particularly in Darwin's recollections of his father, all six foot two and more than three hundred pounds of him, and the father's wisdom in word and deed guided the son throughout life.

"One of his golden rules (a hard one to follow)," said the specter before me, "was 'Never become the friend of any one whom you cannot respect.'" I, too, had been raised with a heavy emphasis on moral character, and Darwin's father was a father to me.

The real fascination of the autobiography, however, lay in Darwin's assessment of his own mind, which history records as one of towering genius. Yet his account stands in such stark opposition to the superficial cleverness, the IQ, which society now takes as genius that my first reaction was denial. No, these could not be the words of a genius. But they were, it's just that Darwin described his mind with the same still honesty that he used in describing any scientific specimen, and what amounts to genius in evolutionary biology happens not to match the popular perception in an age of virtual reality. The druid continued:

> I have no great quickness of wit or apprehension. . . . I am therefore a poor critic: a paper or book, when first read, generally excites my admiration, and it is only after considerable reflection that I perceive the weak points.
>
> My power to follow a long and purely abstract train of thought is very limited.
>
> My memory is extensive, yet hazy; it suffices to make me cautious by vaguely telling me that I have observed or read something opposed to the conclusion which I am drawing . . . and after a time I can generally recollect where to search for my authority. So poor in one sense is my memory, that I

have never been able to remember for more than a few days
a single date or a line of poetry.

I am not very skeptical — a frame of mind which I believe to be injurious to the progress of science; a good deal
of skepticism in a scientific man is advisable to avoid much
loss of time. . . .

I gazed at the specter and remembered how deeply these
observations had affected me, for they matched the ponderous
performance of my own circuitry. I had found a mentor to nurture and protect me among the dazzling university intellects
with whom I was garrisoned.

On the favorable side of the balance I think I am superior to
the common run of men in noticing things which easily escape attention, and in observing them carefully.

As far as I can judge, I am not apt to follow blindly the
lead of other men.

I . . . give up any hypothesis, however much beloved . . .
as soon as facts are shown to be opposed to it . . . for with
the exception of the Coral Reefs I cannot remember a single
first-formed hypothesis which had not after a time to be
given up or greatly modified.

And this bland, unpretentious admission probably had
more practical and philosophic impact on my life than anything
else Darwin wrote. It openly suggested that the mind is a limited
thing, that thinking is groping. From that point onward I had
appreciated that no matter how true an idea may seem at the
moment of revelation, it is nothing more than empty specula-
tion until tested by experimentation or experience. Most notions cannot be tested with scientific rigor, and therefore all of

what we hold right and true — even our most sacred beliefs — should be taken with a skeptical grin, because the court jester is the human mind.

The ghost of Darwin now turned slowly to face me, peering directly into my face. I could not avoid his gaze and stared into his eyes, into his gray-blue eyes, and back in the shadows of his massive brow what I saw was pain. Of all he said in the autobiography, the tormented ending had stayed in my mind, and now these words came back to me, not in the calm, even tones of rational intellect but as a cry in the dark, for this was Darwin contemplating the cost of the scientific life.

My mind seems to have become a kind of machine for grinding laws out of large collections of facts. . . . Up to the age of thirty, or beyond it, poetry of many kinds . . . gave me great pleasure. . . . Pictures gave me considerable, and music very great delight. But now for many years I cannot endure to read a line of poetry: I have tried lately to read Shakespeare and found it so intolerably dull that it nauseated me . . . Music generally sets me thinking too energetically on what I have been at work on . . .

Darwin suffered throughout his life from a strange malady, still undiagnosed, in which such activities as conversing with friends and colleagues brought on nausea and vomiting. The slow erosion of aesthetic sensibilities had struck a private chord of fear in me at the age of thirty, poised as I was on the threshold of a career in science, for I too loved music, poetry, theater, painting — the arts were my home — and if the cost of science was to be the loss of these sensibilities, that cost would be too high.

"The loss of these tastes is a loss of happiness," continued the Arch-Druid, boring into my soul with those anguished, shadowed eyes, "and may possibly be injurious to the intellect, and more probably to the moral character by enfeebling the emotional part of our nature.

"If I had to live my life again I would have made a rule to read some poetry and listen to some music at least once every week; for perhaps the parts of my brain now atrophied could thus have been kept active through use."

Then the apparition turned and slowly tilted his head to gaze down at the cat, who gazed back at him. Each seemed to contemplate the other while I stared at both.

When it comes to supernatural events, I have never been much of a believer, and this séance arose without question from the labyrinths of my brain, an expression of the entire three pounds, of which Hamlet would have said, "There are things in fifteen billion neurons undreamed of in your philosophies, Horatio."

Twice before I had received visitations, and both had changed the direction of my life. They had come through my own voice speaking as if outside and above my conscious being, and they took the form of commandments. The first, at the age of thirteen, informed me that I wanted to be a writer; the second, at the age of thirty-one, declared that my mission as a writer would be to understand the human mind as a biological thing. The second had seemed like a pretty good commandment at the time, a better way to dispose of a Ph.D., at least, than writing novels.

The appearance of Darwin as both cat and man, however, was more like an omen. If an omen, what did it mean? What did any omen mean other than the seer's agenda? Had this truly been Darwin's ghost, his intent would not have been to help me, for no self-respecting spirit would lie still in the grave while someone mucked around with his legacy. He would want to make sure he received credit, possibly royalties. The truth was, this odd little scene in a dormitory room at the University of Bath was simply my mind redefining itself, girding its loins for the long journey ahead, reaffirming the values and qualities of mind that the biologist needs to understand life.

In the Darwin that history reveals to us, I had found a mentor whose values I respected, a mentor whose genius arose from natural events and worked with the thoroughness of dripping water, constructing concepts with the exquisite form and timeless solidity of stalactites, etching down through the layers of time to reveal the structure of the eons. This was a mentor I could envy. Was the genius that understood the forces of nature suited to understand the human mind? Darwin seemed to be saying so.

More to the present point was Darwin's anguish at losing the love of poetry and music and art. "The loss of these tasks is a loss of happiness," rang in my mind " . . . and may possibly be injurious to the intellect, and more probably to the moral character." An alarming prospect, and I concluded it was the Arch-Druid's way — my own mind's way — of urging me to forgo my Western inhibitions and abandon my self to the love of the other Darwin, the Darwin who was probably yowling at that very moment outside my neighbor's door back home in Long Beach.

I looked away momentarily, and when I looked back, both the cat and the man were gone, leaving me to realize against the blank walls of the university how much I missed my big bull's-eye tabby. God did I miss him. "If I had to live my life again, . . ." insisted the Master, "I would set some time aside." Suddenly it was all so clear. Read a little poetry. Love a little cat. Embrace the world, the natural world, the whole world. Find the big perspective. Forget the pissy Englishmen ventilating in the Hooley somewhere in the bowels of the dormitory. A few little insults. A few little people. Big deal. Think about the red teeth and red claws of Darwin's nature. Think about the genius of nature that made those claws and teeth red in the first place. Toughen up, my friend, put the natural perspective to some practical use. Think about the barbaric yawp, that pantheistic celebration of life given us by Mr. Whitman, the poet.

~

I arrived home in the late afternoon and carried my suitcase and bags down the driveway to the flat as if they were no heavier than balloons, the thrust of my anxieties completely neutralizing the weight of physical substance. I thought of only one thing, and my eyes scanned the bushes and nooks where that one thing liked to spend his time. There he was, sitting on the stairs, his orange fur clashing with the ratty red of the outdoor carpeting. I reached down to pet him, and he raised his head, inviting the stroke. When hand touched fur, he stood up, pushed into my hand, and sent joy and relief coursing up my arm into my soul.

My eyes caressed his body, which was covered with dust and looked ragged. There was food in his dish, but it was dried. There was water in his bowl, but dust lay on the surface. It was clear that Berdy had not attended to his needs the way I would have done, but all in all Darwin seemed in good shape. He preceded me up the stairs and scooted into *his* flat the instant I opened the door.

We ate like kings that night, particularly Darwin; I suspect it was the first time he had eaten canned salmon and Trader Joe's duck pâté. Then it came time for bed, and this time I had no intention of putting this creature outdoors for the night. From now on he could stay inside. No sooner had I turned off the light than a slender thread of sound began to twist in the darkness, and as it twisted, the sound swelled. And swelled. My perceptions were caught completely off guard; my first thought was the local air-raid siren. The emergency broadcast system. My second thought was less hysterical, but only slightly. Welcome home, I thought. Now the cat wanted to go *out* for the night.

I got up, gave him some more pâté, and went back to bed. My lights were out before I hit the mattress and sank down, down into the comatose depths of the fossil record.

5   NUPTIALS

ROCKSSHITOHGODBREAKWATERCRASH
KICKOUT STOPROCKS (convulsion kick).

Find night light Gropewhack Unnhh!!! Nolamp . . . Wrongside . . .

OhJezus — on bed!! . . . Something on bed! Mattress sinks underit. Toward me! . . . Thinkthink.

THE CAT!!!

Cat has jumped on bed. . . . At me. Scalp crawls. Mind coming online. Cat Jump — Cat Thump: Circuit shock . . . Not in boat, inebriated, not drifting at rocks . . .

Cat coming closer. Something presses on right thigh . . . a paw. Two paws. Fifteen pounds on two small feet. Like high-heeled shoes, pressing sharply into flesh. Climbing onto both thighs, circles around, plops heavily on side atop me.

I roll abruptly to the right under the sheets. Cat slides from my legs and lodges inertly against me. Am now groggily awake. We lie motionless for several moments, while I consider what to do.

Here I am, flat on my back. I have broken through the barriers of my culture and my private feelings and accepted an animal into my circle of loved ones. I am jet-lagged. I am almost delirious with fatigue. Yes, I feel true affection for this little creature. No, I do not want to sleep with him.

I reach out wearily and begin to stroke the furry lump, feeling very gingerly for the head and mouth. No reaction. No purring. The ball is in my court and Darwin, apparently, has every intention of keeping it there. I nudge him gently toward the edge of the bed. He relaxes and lies limp, letting his body bend to absorb my efforts without moving. I reach out with both hands and try to scoot him along. This works for about a foot and a half, when Darwin suddenly seems to gain about fifty pounds and cannot be budged. I push harder. No movement. What the . . . ?

I turn on the light and there he lies, ears laid back, claws hooked deep into the mattress with knuckles arched. He has no intention of abandoning our bed. What to do? I am in no shape to do anything, so I do the inevitable and doze off.

Some time later, I think, a dream turns bad, I think. I lie on a Caribbean beach with marimba music and the scent of tropical flowers wafting on the breeze. The sun soaks into my skin and I savor the feel of the hot sand on my palms . . . but the hot sand grows hotter . . . hotter . . . and becomes a stabbing pain in my left hand. At which point I wake up to find my hand clamped firmly in Darwin's mouth; I cannot *see* my hand — the lights are off — but I don't need lights. In my mind's eye I clearly see a hand sandwich.

Yeow! I try to pull back, but Darwin won't let me off so easily. I jerk back with a terrified reflex and finally extract myself from his jaws. He must have moved while I slept and bedded down opposite my chest. In the course of dreaming, I laid my hand on Darwin's head, oblivious to the sin I was committing, and he is letting me know it was a mortal one.

With expletives resonating through the flat, I run to the kitchen to check my hand for puncture wounds. I find only white marks, Darwin's way of reminding me it could have been much worse. Again I go back to bed and fall directly into a coma.

The next morning I regain consciousness around nine o'clock. I lie on my back for a while, assembling the various pieces of my mind, and look over at Darwin, who lies next to me, stretched out full length. How relaxed he looks, how innocent in the privacy of sleep. My eyes run over his body, feeling his rich colors and lingering on his elegant markings, and I am struck by his size. From nose to tip of tail he looks about four feet long. He lies on his side with his legs reaching toward me as if wanting to embrace, and something in the curve of his wrists and the innocence in his slumbering sprawl summons up the fairer sex and memories of a few with whom I have shared intimacies. The thought occurs that these young women, too, have memories of the fouler sex, of which I am a shining example, in similar postures of affection and trust. At which point my inner voice roars out, "My God!! — you're in a relationship with a cat!!"

Ah yes, the morning after. There I lie in the sinful aftermath of having slept with a beast, and I feel not a twinge of guilt. On the contrary, I am glowing with the warmth of a first-time husband, for my soul has been liberated and I have no more reservations. I intend to throw myself into this relationship. Not even the eggs and castings of all the fleas who have shared our bed can dampen my spirit.

Thus Darwin and I became man and cat.

6) Honeymoon
Prognosis

IN THE FOLLOWING WEEKS I bought Darwin expensive
specialty foods — nothing but the best for my small friend. He
responded by getting fat, eventually compressing the scales at
nearly fifteen and a half pounds. In an age when leanness has
become something of a medical religion, I felt some concern
over this gain in weight, but it gave me such pleasure to watch
him revel in his meals that I could not resist giving him large
helpings. This in turn allowed me to enter the mindset of Peter
Paul Rubens and other Baroque painters: for the first time I understood their tastes in hefty women. A person of weight was a
prosperous person, a symbol of power, particularly in an age
when good food was a privilege, and Rubens, consummate poli-
tician of the royal courts, understood perfectly how to butter his
bread.

Nevertheless, it seemed wise, given my newfound responsi-
bilities, to see what the medical experts said about weight gain
and other matters of health, so I bought a veterinary handbook
for reference. What I found on its cold white pages was a gallery
of horrors, illustrated with weeping wounds and bulging tumors and pus-rimmed eyes. There were failing hearts and dam-

aged kidneys and inflamed bowels and allergic seizures. There were fleas and ticks and mange mites and tapeworms and hookworms and liver flukes. Then of course there were the microbes, including a funereal list of viruses with names like FIP, FPV, FIV, and FeLV (the only omission was RIP). These abbreviations were derived from the medical names: feline infectious peritonitis, feline panleukopenia virus, feline immunodeficiency virus, and feline leukemia virus — a list of doom revealing a dark world of feline pathology that ran exactly parallel to our own, including two retroviruses, FIV and FeLV, that acted much like the retrovirus we humans know as HIV. I didn't dwell on the details, but according to the book, both viruses attacked the white blood cells, particularly the T-cells essential to the immune system; both caused the victim to lose weight and waste away; and both led to all sorts of secondary diseases, including cancers like lymphoma and leukemia.

I shelved that book, hoping never again to read such a horrible tract and thinking that if cats had nine lives, they had got the short end of the stick. Nine were not nearly enough. Nine lives, however, was a value judgment made by those who have no particular love of cats or who downright dislike them. To those of this persuasion, the cat is a tough, resilient survivor, whose ancestry goes back sixty million years. But when you fall in love with a cat, you take away eight of those lives, leaving it with only one and creating a small, frail creature with little chance of living out a long, healthy life.

What could I possibly conclude after reading this veterinary tract but that Darwin had to be vaccinated? He also had to have his teeth cleaned. The substance caking his teeth was tartar, and the red line that ran along the base of his teeth was gingivitis, an infection of the gums brought on by tartar buildup. The infection would spread to the jawbone, eventually loosen his teeth, and cause their loss.

We had returned to the matter of finding a good veterinarian. This time there was no turning back and no time to wait. As Darwin's guardian it was my duty and my loving compulsion to guard his health. But in order to find this ideal veterinarian, Fate would have to intervene.

About a week later, I conducted a telephone interview with a woman named Bonnie Mader for a column I was writing on the human-animal bond. Bonnie was a professional grief counselor and had founded the first pet-loss hotline at the University of California, Davis, School of Veterinary Medicine. The service had proven so popular that it could not attend to all the requests, and as I worked into the interview I learned from Bonnie that some people, on losing their animal companions, commit suicide or seriously contemplate it. I felt a strong connection with Darwin, but this seemed a bit much. What else could you conclude but that these poor souls were emotional misfits, citizens of the psychological fringe?

Even so, you could not so easily dismiss the enormous number of people, cloaked in the anonymity of a telephone conversation, who called for help during their moments of deepest grief. Bonnie Mader had struck a nerve; out there in the vast closet of American society lived multitudes who suffered in silence, fearing ridicule and derision, ashamed to admit they had developed deep bonds with simple creatures.

We talked for an hour that day, and as we were about to hang up, I happened to mention where I lived.

"Oh," said Bonnie, "my brother just bought a veterinary practice in Long Beach."

"You're kidding me. *I* live in Long Beach — whereabouts is it?"

Bonnie went on to explain that her brother now owned the Long Beach Animal Hospital — apparently, when a newly emerged veterinarian looks to buy a practice, he chooses what the market has to offer, wherever that may be. So Dr. Doug Mader had recently moved to this city of 450,000 citizens just south of Los Angeles, a city that used to be called Iowa by the Sea in honor of the immigrants who came during the Depression. Sometimes it was called the Home of the Newly Wed and the Nearly Dead, with reference to the young families just starting out and the large number of old folks awaiting their ultimate destinations. If this were not reason enough to settle in Long Beach, there was also the *Queen Mary,* permanently docked and rusting in the harbor, and next to her, beneath an enormous geodesic dome, the *Spruce Goose,* built by Howard Hughes at the end of World War II. Long Beach boasted the narrowest house in the nation as well as the largest United States flag. If Doug Mader needed any further inducement, Long Beach was the home of Darwin.

The next day I took Darwin, howling and urinating in his new transport box, to meet Bonnie's brother.

Dr. Mader was an intense, lean, dark-haired man in his early thirties who had been nicknamed "Worker Bee" in vet school for his implacable energy. According to Bonnie, his primary motivation was actually to help animals, and he sponsored a pro bono program to save and rehabilitate injured wildlife. He had even set up a student intern program with U.C. Davis; consequently, he had close ties to the university and access to the latest in veterinary research and medicine. Darwin, his eyes dilated black, did not appreciate the significance of this, but he was soon to be in the best hands that medicine could hold forth.

Dr. Mader laid those hands on Darwin and lifted him deftly from his box. He laid him gently on the examination table. Darwin seemed submissive, as if acknowledging the hands of a higher power, and lay dutifully on his side while the doctor pressed a stethoscope to his ribs and listened to his heart. He complied without resistance as the doctor prodded and massaged his abdomen for tumors and other telltale symptoms listed in the book of veterinary horrors. Then, declaring that Darwin's teeth needed cleaning, Dr. Mader explained that this required a blood test.

Everything leaves evidence in the blood, a fact that I, as a lapsed biologist, understood. If Darwin's kidneys were damaged, that would alter the nitrogen profile. If he had parasites or any number of diseases, they would activate the immune system and induce changes in the blood cells; if liver disease, enzyme changes; and so on. The first order of business on starting a cat on regular medical care was to test its blood. This would establish a record against which future tests could be compared.

Finally the exam ended. The results would come back the next day, and then we would schedule the cleaning. Dr. Mader had mentioned nothing about a heart murmur, which corroborated my suspicions about Dr. Grog.

At eight A.M. the phone rang, waking me from the sleep of the late-rising writer. "This is Dr. Mader" said a flat, matter-of-fact voice. "We got Darwin's blood work back. We have good news and we have bad news. I'll give you the bad news first. Darwin has the feline leukemia virus."

My mind ceased to think. I sat there on the side of the bed and my eyes cast about aimlessly, noticing the wrinkles and folds in the bedspread, the bumps on the stucco wall. A sensation similar to chills spread over my shoulders, bringing back that summer night so many years before when I had just come home from visiting a friend. As I parked the car, my mother ran from the house to greet me with a look I had never seen before. "Something terrible's happened," she said with a strange, hoarse severity in her voice. "There's been an accident. Your grandfather was killed. Your grandmother is critical . . ."

That same feeling of horror and shock swept over me as I gazed blindly at the walls, my mind frozen in its frame, suspended in the nothingness of loss. All the neurons, all the synapses, all the enzymes and ions and other chemicals of the brain devoted to the memories and thoughts of Darwin would soon have no further use. At that existential instant I somehow knew that Darwin would not have much time but that it was going to be intense time. It was going to burn.

The doctor continued to talk, his words bleeding together in a background mumble while my self blundered aimlessly this way and that. Then the words began to break free of the mumble and enter my mind.

" . . . the good news is that we don't know how fast the disease will move. In fact, some cats seem to live with the virus and never come down with symptoms. With others it goes so slowly that you can enjoy them for months, even years. We have to make a decision, though."

I tried to ask what sort of decision, but tears welled up and my throat wouldn't work. I paused, forced composure on myself, and managed a steady, low voice.

"A decision? What kind of decision?"

"Do you want to go ahead with the teeth, or put him down?" said Mader's flat, dispassionate voice.

"Put him *down!?*" I was incredulous.

"Yes — many people decide to do that when FeLV is diagnosed."

"But . . . *Why?* Don't a lot of cats survive it? Why give up without seeing what happens?"

"Some owners would rather end it now on a positive note and save their pets from illness."

I had to decide, and I had to decide now . . . but we were talking about Darwin's *life.* At times like this, when profound denials collide like tectonic plates deep in the guts, my mind sometimes separates into pure reason and pure hysteria. The rational self rises into the air and peers down with detached fascination as my emotions writhe on the hook of life.

"Why spend $160 we can't afford to clean the teeth of a dying cat?" said my rational side. These were quicksilver thoughts, almost instantaneous, more feeling than conscious, but there was no denying the calculations. I knew from the veterinary book that roughly 70 percent of infected cats die in a relatively short time after being diagnosed with the feline leukemia virus, and seven in ten was poor odds for investing money on clean teeth.

Hope, however, was as tough and strong as it was illogical. Odds are a measure of group history and could not reveal the course of Darwin's individual fate. Viewed with hope, 30 percent of FeLV positive cats seemed to tolerate the microbe — and three chances in ten became fabulous odds. My mouth opened, and off in the distance I heard my voice say, "Yes . . . Yes — let's go ahead with the cleaning."

7 HOPE, INTIMACY, JEALOUSY

THE HOSPITAL called around three o'clock, and I brought
Darwin home with clean teeth and a hangover. Still affected by
the anesthetic, he wobbled about when I lifted him from his
transport box and placed him on the floor, then sat down to get
his bearings, while I stood, looking down at him. *He would not
always be there.* My rational mind grasped the notion without
blinking, but only now were my deeper feelings beginning to
comprehend. I reached down, picked Darwin up, clutched him
to my chest, pressed my cheek to the side of his head, closed my
eyes, and stood there rocking back and forth while his body
warmth flowed into me.

A collage of images passed before my eyes, creatures I had
killed as a young hunter and later as a scientist. A mallard drake:
I spread his limp wings to admire the shimmering blues and
greens of his speculum, stroked the iridescent green of his limp
neck, looked into the clear blankness of eyes whose soul I had
just snuffed out. A laboratory rat sacrificed on the altar of teaching, its skull bashed against the sink; I cut through the stomach
skin to extract its liver, noting the elegant packing of stomach,

spleen, kidney, intestines, feeling the wet warmth. A pheasant, a sunfish, a jackrabbit, a gopher snake — on and on in a progression of death.

I thought of how the intellect had reduced these creatures to nothing more than teaching devices, divine perhaps, sublime in form and function, inspired in the aesthetics of color, texture, shape, miraculous in the chemistry and physiology of living substance, but devices nonetheless. Some sort of denial had, apparently, prevented the emotions of empathy and compassion from burning through. I recalled how it felt to regard these lifeless remains with the cold blue flame of intellectual curiosity, and I shuddered.

Again I pressed Darwin's furry softness against my skin, clutching this warm, supple miracle of existence, and rocked back and forth in a rhythm of gratitude and grief for whatever time we might have left.

As for Darwin himself, after a few minutes of my maudlin reveries he began to struggle and squirm, wanting only to be set down. He doddered around from the fading spell of anesthesia, found a private place beneath the couch, curled up, and went to sleep. A few hours later he was meowing loudly at the door, demanding unsuccessfully to go outside. So far as he was concerned, it was back to the business of life with its ceaseless vigilance and metronomic border patrols.

⁓

For me, however, this was not the Darwin I had taken to the vet. This was the Darwin that medical opinion said was dying; I could no longer gaze upon him in innocence, but found myself looking for indications that the dreadful progression had begun. The virus had infected my mind just as surely as it had infected Darwin's body. Medical knowledge does that. It sets the mind against itself in a self-generating cycle of paranoia, each fear fueling the next. It takes bliss and makes despair. I had assumed we had all the time in the world, years and years, before death parted our ways, and to realize we had little time left is the kind of revelation that seeps slowly into the molecules and neurons of the brain. It made a miser of me. It transformed time into a necklace of precious minutes, hours, days, and I was going to hoard them like diamonds.

But above all, this awareness of Darwin's doom started a war of mental attrition that would carry on to the end. On one side was the cold blue reason that measures reality and calculates odds, that weighs value and worth — the computer mind that had voted to end Darwin's life instead of cleaning his teeth. On the other side was the mind of hope, which twists, bends, deletes, and otherwise alters reality to contrive happy endings and is by its very nature incompatible with truth.

The days passed with no signs of disease, and hope grew stronger. I felt more and more relieved because we had, apparently, ducked the executioner's ax. We had beaten the 70 percent odds that favored the progression of the virus. The calculating mind pointed out that it was too soon to draw such conclusions, that the game was not over, but hope would have none of it and flooded my spirits with the prospect of Darwin's recovery.

And so, driven by love and gratitude for the good fortune of having Darwin in my home, I observed him with the magnified resolution of a microscope. Each motion, each gesture, each step, each grooming lick, for anxiety, it seems, raises awareness, sharpens the eyes, nose, and ears, sensitizes the tactile skin. I noticed things that scientific-intellectual curiosity overlooked completely, and I began to see how much larger and more complex Darwin's mind was than I had ever suspected.

We evolved a new routine in which Darwin spent the nights indoors. In fact, he had a curfew: no dinner until the door had closed behind him. In the morning, following breakfast, he would walk over and rub against my legs, looking up to catch my eye. When our eyes locked, he would turn and trot purposefully toward the door with his ears swiveled back to hear if I was following. Then he would stop and stare at the door in an unmistakable gesture that he wanted out. If I ignored him, an expanding yowl rose into the air like escaping steam and continued, rising and falling, until I obeyed his command.

His activity pattern, on the other hand, was continuously changing. Most creatures fall, like humans, into variations of the morning and evening commute. When I was a boy, the big event of dusk was watching the quail fly to roost in the cypress and the eucalyptus trees. The event of the dawn, at least in the spring, was the red-shafted flicker hammering his challenge to the world on our rain gutters, transforming the house into a snare drum. You could set your clocks by these events; they illustrate the circadian rhythms in which most living things play out their lives.

Darwin, though not exempt from daily rhythms, did not express such tightly wound patterns. This was most evident in his daily patrols, which I could never predict. He never walked his paths at the same time or on consecutive days. So with his sleeping places. Every week or so, he would choose a new place to bed down. Sometimes he would prowl the flat off and on during the entire night, pleading to go outside. The next night he would sink into a comatose sleep and would not beg to go outdoors until I arose the next morning.

With observations like these I reverted to the thinking of biology and contemplated the causes that lie beneath personal experience — causes beneath conscious control — that make us what we are. Because I had seen a similar randomness in the activities of all the cats I had watched from my second-floor vantage point, I concluded that it was probably the innate trait of a predator, an adaptation to keep the prey guessing, unable to predict where the next attack will occur. Darwin made evolutionary sense.

I also began to recognize the signature of neoteny on Darwin's habits. Neoteny denotes the retention of juvenile traits in adulthood, and whereas many of us have been accused of this condition in our personal behavior, as a species we humans are said to be neotenic apes because in our adult form we exhibit physical traits found in juvenile apes. A case in point is the thinness of the human skull and its delicate construction, which resembles that of a young chimpanzee. In contrast to this stand the massive cranial ridges and crests and thick plates of the adult ape.

Neoteny also applies to behavior and implies a genetic foundation. As Darwin and I became more intimate, I began to recognize that much of his behavior was that of a kitten. When his back was stroked he would reach out and knead the rug with his toes and claws, just as a kitten kneads its mother's breast when nursing. In fact, many of the postures and gestures I found so appealing were the behavior of a kitten manipulating its mother. After a while I saw these signs everywhere, in the way he begged for food, trying to lock eyes and penetrate my soul, in the way he rubbed against my legs when wanting something, in the mewling and crying that sounded so similar to that of the human infant.

A wild cat, on the other hand, would behave very differently. Had Darwin been born of wild parents, the ancestral *Felis sylvestris,* he would have become much more skittish and highstrung. His spirit would stand aloof, unto itself, there would be no begging for favors and no asking for quarter, and when provoked he would lash out ferociously in self-defense. In other words, he would have grown up to become a true adult. He would have acquired the toughness and aggressiveness to survive on his own, keep a territory, win access to females in heat, and he would be very difficult, if not impossible, to control. By the genetics of a wild nature, he would be ill suited for human companionship, even dangerous.

For a pet, the average person needs an animal that remains a child. Children, raised with traditional discipline, can be made to submit — humor me on this — forced to take direction and accept training, molded to the will of the adult. Is that not the essence of rearing children, to produce a fine, god-fearing adult with good values and good character, in the parent's image? So nature gives us a window of opportunity to shape and mold, before the child becomes a large, hairy adult with nose rings, full-body tattoos, and the defiance, strength, and obstinacy to resist domination. Oh, the neoteny of it all. Just as wild kittens grow up to become wild cats, so children grow up to become human beings. Perhaps the fundamental goal of all domestication is to draw this genetic compliance of childhood forward and fix it in the adult animal, creating a creature that we can bend to our will and our emotional needs.

One day I walked into the living room and saw Darwin sitting on the floor with his back toward me, a furry little pear of a figure, with ears laid back to monitor my movement and this fuzzy, animated snake of a tail protruding from his rump and twitching each time I said "Darwin." Affection and sadness mixed with gratitude for his presence, and without warning, "Darwin . . . Little Gumbie" came from my mouth. When his tail twitched again, out came "Little Bumbie Buddy," followed by "Muffin Buffin."

Muffin Buffin? I had always regarded little old ladies who baby-babbled at their Poopy-Doo Poodles and Itty Bitty Kitties through a youthful lens of pity and condescension, incapable of grasping how it would feel to join their company. Now that I had, it bothered me not one whit. All conceit had been stripped away, and I leaned into Darwin's affection the way a cat leans into a stroke. Reduced to little more than my most private intimacy, I realized with something akin to shock that my spirit was neither different from nor superior to the spirit of my small companion. Our souls stood together in absolute equality. Equality implies identity. A feeling of camaraderie arose from the deep regions of my brain and spread over my skin like a blush.

It triggered a creativity that would best be mentioned with a blush. "Teapot," "Teacup," "Darwin Rex," "Little Furball," "Pumpkin" — over the following weeks the variations came in infinite profusion, and Darwin's tail kept twitching, probably more from irritation than from any recognition of his name.

And what followed from intimacy? One night a lady of my friendly acquaintance dropped by, and, having enjoyed a fine meal, we turned to intimate talk and good wine. The candles flickering with warmth, we sipped and gazed with anticipation into each other's eyes. Suddenly, with shattering glasses and falling bottles, Darwin exploded onto the table, squarely between us. He sat down, folded his tail around his forepaws, calmly placed first the right then the left on top of it, and stared inscrutably with those orange stiletto eyes.

Jealousy followed from intimacy. Did Darwin want exclusive rights to my innermost privacy? That was difficult to know, but I could not deny the possibility. Some odd incidents over the past few weeks then began to make sense, and I understood why he had been jumping onto the leather armchair, the forbidden land, when he knew full well that a stream of water would soon shoot from the heavens and nail him. Invariably he challenged me while I talked on the telephone, and I had assumed he was calculating that I could not drop the receiver and reach the spray bottle in time to enforce the law. I had also assumed that he did this for the sheer pleasure of getting his way and defying my authority. Now I realized that it was a matter of feeling neglected, hence a matter of jealousy, for the telephone was consuming my attention and shutting him out.

With weeks having passed and Darwin still showing no signs of illness, hope was strutting its stuff and life had pretty much returned to normal. Therefore he could use some training, I decided. It seemed only natural to teach him some culture, some manners, demand a little self-restraint.

The first project was trivial and idle and had nothing to do with civil behavior. I resolved to train Darwin to bark for his dinner like a small dog. It would amuse my friends, it would amuse me. The thought did not occur to me that it might not amuse Darwin.

We began with the standard performance-and-reward

method of training. I always made sure Darwin was hungry, consequently eager to do whatever was necessary to acquire food. With a broad flourish I would place a can of food in the opener and crank slowly around the perimeter until the blade came full circle and severed the lid. I would blow across the top of the can and waft the aroma of delicious cat food in his direction. Then, with slow, broad movements and loud clinking of spoon against dish, I would make a show of preparing his dinner. I would talk to him, modulating my voice with exaggerated emphases. "Oh. . . . BOY!!!! The BIG MOMENT!!! The MOMENT you've been WAITING for!!!

Darwin would sit at my feet, staring intently upward at elbows and jowls, periodically standing and circling impatiently, then sitting down again. Sometimes he would stand, wrap his tail around my calf, and embrace me for ten or fifteen seconds. Sometimes he would meow. Sometimes he would bark in that little, faraway voice. Immediately I would place the bowl in front of him so he would associate his barking with the delivery of food. If he did not bark, the next step was to brandish the food over his head and cajole. Sometimes it worked, often it didn't. Whatever the reasons, it soon became clear that Darwin's mind was not a simple Skinner box — that bizarre container of conditioned responses laid on the innocent world by the psychologist B. F. Skinner, who thought that all behavior was constructed top to bottom from learning and conditioned reflexes and had no genetic aspects. Darwin, however, would not enter that mythical box. As I persisted in my attempts to train him, I got the distinct impression that if he could speak, he'd be muttering dark comments about Big F Skinner during our sessions.

Well, if I could not mold Darwin into a walking, barking

cat, at least I could teach him some civil restraint, and I returned to the project, abandoned earlier, of combing his hindquarters for fleas, where many went to spend their vacations. Now that we had arrived at a state of spiritual intimacy, surely he would be receptive. As before, I waited until he was eating, then tried ever so gently to run the comb through the thick fur on the backside of his hind legs. I have an image — I will die with it — of Darwin, ears laid back, pupils dilated, legs splayed, claws digging into the linoleum floor, tail snapping back and forth, and a deep, hoarse yowl coming from his throat, the posture and vocal emissions of a cat about to kill.

Apparently I had no choice but to back off. I did so, but sanctimoniously. After all, I was only doing this for him.

8 FRIENDSHIP AND EQUALITY

I NOW REALIZED that Darwin's feelings had to be considered. He had become my constant companion, my friend, and in its noblest form, friendship requires equality, that Holy Grail of Western civilization. The American colonies, in declaring themselves independent of England, even went so far as to contend that "all men are created equal." A civil war, a hundred years of resentment, and a less-than-civil movement later, the concept of "men" was remodeled to include all human beings. And what has this to do with Darwin?

It has to do with the peck order, or dominance hierarchy, first noted in the behavior of the chicken, and taken to supreme complexity and refinement in the human being. The peck order has its roots in the limbic system of the brain, also known as the reptilian complex. In the course of evolution, the mammals arose from the reptiles, and as the mammals evolved, the cerebrum region at the front (or top) of the reptilian brain expanded and expanded and folded back over itself, forming another layer that completely covered the section from which it arose. That layer, the cerebrum, now envelops the older reptilian brain, so inside the mammalian brain lies the reptilian complex, literally, an inner lizard.

The cerebrum gives us reason and thought and language and awareness of the self. It is Hamlet's cerebrum that contemplates Yorick. The reptilian complex gives us our appetites and lusts and sexual drives; it also gives us our competitive urges and our aggressions and fears. It is the inner lizard that drives us to participate in dominance hierarchies, rising up huge and scaly in the minds of us all, clothed in the barest bikini of rational thought and gesturing with its middle digit at political correctness.

What we conceive as equality is actually a special expression of the peck order — or, rather, its suppression. In social creatures like wolves, lions, hyenas, monkeys, mandrills, baboons, chimpanzees, gorillas, and on throughout the vertebrate world, groups are structured more or less according to rank, usually with some form of top dog, or alpha, to which all others defer most of the time. Beneath the alpha lurk the betas, which defer to the alpha but not to others, and so on down the ladder of dominance, finally descending to an omega creature which defers to all and which all others bully, harass, disrespect.

In human societies, those on the lower rungs of the ladder find the experience most unpleasant, and the civilizations of the West are dedicated to suppressing the more physical and brutal methods of asserting dominance. And that essentially is what Mr. Jefferson addressed in his declaration that all men are created equal, that they have the right to life, liberty, and the pursuit of happiness. In order to achieve these ends, it is necessary to create the laws of civil and human rights so that we are all, in the ideal state, crowded as equals onto the same rung of the social ladder.

Which brings us, finally, to the topic of friendship and equality. Bearing in mind that personal interactions lead to peck orders, but that friendship implies equality, it follows that you cannot pull rank on friends. You cannot bully friends, because to do so is to put them down, to place them on a lower rung and render them inferior. If friends do not agree with your position, you must restrain the urge to force your will upon them. All humans contain the most sensitive put-down sensors, which detect the slightest challenge to one's rank and to one's self-esteem; self-esteem is fundamentally linked with how successfully one defends one's rank.

I didn't like Darwin's refusal to grant combing rights, but if I valued him as a friend, I had no choice but to acquiesce. Oblivious to their good fortune, the fleas continued to prosper and multiply on Darwin's rump.

~

There was, however, one exception to all this equality and consideration in our relationship, and that concerned the good old-fashioned practice of teasing. Now teasing is an interesting activity because, reduced to its fundamental essence, one party dominates the other, as with tickling, and for the duration of the bout the relationship is decidedly unequal. Teasing can be malicious, as with schoolyard bullies, but within any good relationship it is usually an expression of affection, even love, the aggressor respecting limits and stopping short of pain.

Consider it from the reverse point of view: whom do you *not* tease? You do not tease the king. You do not even *think* about teasing the queen. Nor the Mafia don, nor the gangbanger, nor the police officer. You do not tease those you fear, those with power greater than your own.

Nevertheless, presuming a relationship grounded in love and respect, it can reasonably be said that teasing is part of a normal, happy childhood and leaves one with fond memories, the weaker trusting the stronger to control his strength in the name of love. The father tickles the child; the husband tickles the wife; the older sibling tickles the younger; and universally, children tease the dog, the cat, the hamster, whatever is available on the lower rungs.

Over the course of my boyhood we had two dogs, an Airedale named Duchess and later, a Doberman-shepherd mix named King, and one of my enduring pleasures was now and then to inflict irritation in a lighthearted way. This was nothing less than a fledgling act of human dominance, for I was learning the trade of my species. An essential aspect of that trade is learning to square one's deeds with one's self-image and emerge as an angel. Rationalization is the great genius of the human creature. Even as a young boy I was a prodigy.

I was six or seven when Duchess arrived as a little bundle of kinky, wiry hair, and already I was quite capable of rationalizing the act of teasing. I was just having a little fun. I never wanted to hurt my puppy. The thought of causing physical pain, much less injury, horrified me. I teased in the spirit of affection. Teasing was a bonding experience; it implied an intimacy that one could not have with a strange dog, and in my unbiased opinion, Duchess enjoyed it as much as I did.

A relationship with dogs, far more than with cats, is based on subordination, not equality. It is a master/subject relationship, and there is no helping it. The dog will not have it any other way. Its mind has been calibrated to exist within the structure of a pack, and the pack functions as a team, a predatory and domestic machine. Within which rank and role cannot be separated and are continually changing in relation to age, health, social situation. The dog depends for survival on its ability to adjust to the moods and needs of its pack mates and is highly sensitive to them.

And so we peer down from our intellectual height upon that writhing, licking, yapping, quivering, grinning, cringing, salivating, yelping, urinating display of appeasement gestures known as man's best friend. It is like the wristwatch that takes a lickin' and keeps on tickin', except that the dog takes a kickin' and keeps on lickin'. And what do we see? Why, we see terms of endearment, of course. We see a creature expressing its unwavering loyalty, its unquestioning acceptance, its complete forgiveness, its unequivocal love. Whether we like it or not, we are supremely dominant in this relationship and, though most of us would never recognize or admit it if we did, dominance makes us feel very, very affectionate.

What drives the dog to such incontinence? The answer lies in the behavior of lower-ranking wolves as they address wolves of higher rank. There you see that these appeasements are not expressions of euphoric friendship, loyalty, character, and other human projections. No, these are expressions of the most desperate anxiety. This is primal supplication to the pack's good graces, for which the dog will endure any insult and accept the lowest rank.

Not to belittle the dog. Who cannot love this earnest, innocent creature? Compassion is more appropriate than ridicule in contemplating its excesses, as it is in contemplating the excesses of us all. Like any creature, the dog thinks and feels in the manner its brain is constructed to *make* it think and feel; in the natural overview, its behavior is no fault of its own.

The brain and mind of the dog are fashioned for politics — the pursuit of one's agenda through the exercise of power and skill in the society of one's own kind. Politics is the inevitable upshot of group living, because to live in a group is to give up the territory one would have as a solitary creature. Another way of seeing it is to imagine the territory of each member consolidated into one large territory that must be shared. The individual territories are, in a sense, stacked up, one upon the other, and this forms a hierarchy, a dominance hierarchy, and where you sit on this structure depends on how well you ply your political strengths and skills. The dog is therefore similar in its sociopolitical orientation to that of its human master, and its behavior speaks naturally to our gregarious emotions.

The cat, however, is an interesting case because its basic nature is solitary, yet it too fits the human mind. The solitary animal is usually a territorial animal, so most encounters with other cats are militant. Confrontation produces the basic emotions of threat and violence, and these in turn inspire broad gestures, strong motions: the snarl, the long, virtuoso yowl, the lashing tail, the laid-back ears, the arched back and expanded tail. Aside from that, the cat in nature has little need to express itself, particularly in facial gestures, for it has no one to face and no one to communicate with. Consequently it lacks the equipment to display the mercurial nuances of feeling and mood that distinguish the dog. The machinery is simply not there.

This is not to say that cats cannot communicate peacefully with one another, face to face, for clearly they can. But communication is simple and broad, like ritualized grooming or sniffing of the anal glands, and usually serves to identify the individuals and help establish dominance or subordination. Such gestures of action and posture and olfaction have probably remained in place from infancy and kittenhood, the brief interlude when cats are social creatures and must interact with mother and siblings. It is a relatively crude and rudimentary suite of behaviors compared to the nuanced communication among dogs; and it is the consequence of solitary life. That is why the cat simply stares and stares at its human benefactor. The human may frown back, he may smile or grimace; he may clown and mug and act the complete fool, and the cat, without the machinery to respond, stares on, without expression.

And that is why it finds a home in the human mind: the cat relieves the solitude of the self, for the self, sealed in the bell jar of the skull, is in a state of solitary confinement. The cat is a kindred spirit to the private, ruminating side of our mind, and it slips unobtrusively in and out of our solitude as it will. The cat draws us into contemplation and introspection. By its nature, the cat respects the privacy of the mind, and in this deference, reveals how invasive is the mind of another human — "What are you *thinking?!!!*" "None of your damned business!!!" In the asocial nature of the cat we find the deep, silent pleasure of simply being in the presence of another living thing, communion.

The social nature of the dog, on the other hand, brings with it not only the behavior to communicate its fluttering emotions and sliding moods, but also the mental equipment to read the signals. The brain of the dog, like the brain of the human, is constructed to look for sophisticated visual cues in the face. In this it is fundamentally different from the cat.

To illustrate, consider what happens if you confront your dog with *no* facial signals, at the same time placing it in a high-stress quandary by simply staring. This is perceived as a challenge, an act of aggression. (The eye challenge also exists among humans and can provoke an aggressive reaction, particularly from strangers.) The dog scans your face, compulsively looking for cues. You continue to stare. The dog tosses its head. It yaps plaintively, imploringly. Then it drops its forequarters to the ground, and with its rear still raised, attempts to draw you into play, defusing the confrontation. Motionless, expressionless, you stare on. The dog lies down and rolls on its back in a posture of submission, beside itself with anxiety at the inexplicable aggression from a dominant pack member who refuses to communicate.

The dog, then, draws us out of the self and into the social world of action and reaction, allowing us to forget the fears and doubts and cares and sorrows of solitary confinement and throw ourselves into the rough-and-tumble competition of a hierarchical life.

A proper dog is therefore good with children. It will take their abuse with its good, subservient nature and not strike back. Which makes this affable creature a fine subject on which a young human can practice its humanity.

⌐⌐

Which brings us back to the topic of friendship, equality, and teasing. A quality tease had to meet certain criteria. It could not inflict injury or pain. Irritation, yes, of course, but no pain. It had to be clever, a challenge to the "master's" ingenuity, but friendly teasing was intended only to exasperate, and any sign of real anger brought an immediate halt to the campaign.

An example of my ingenuity was to feed King or Duchess a tablespoon of peanut butter. Incredulous at its good fortune, either dog would lap the stuff from the spoon, then spend the next fifteen minutes with head cocked to one side or the other while trying to dredge the oily gob from the roof of its mouth with the back of the tongue while I howled with glee in my superior intellect.

My greatest triumph occurred when I was seven years of age and had moved with my parents to an old, rundown farm near Doylestown in eastern Pennsylvania. One day, shortly after we had moved in, I was exploring near the barn while Duchess was snuffling about for odors in the grass, holding her short, stiff tail in the air like a flagstaff. I found a short piece of rope, held it in my hand, and stared. Then my gaze traveled to Duchess's tail. And then, naturally, came the inspiration. Why not tie the rope to the tail? Why not? No further reason was needed.

I wrapped the rope around her tail several times and secured it with a simple knot. Looking at my handiwork, my miniature human mind noticed the other end of the rope. Why not tie it to Duchess's hind leg? So I did. I did because I could. Just as I finished the second knot, Duchess took a small step, which drew the rope taut and jerked her tail. This triggered a reflexive leap which nearly yanked her tail from its socket. The first leap triggered a second, which triggered a third, and Duchess disappeared over the horizon as yelps of terror came wafting back.

Why do such events strike us as funny? Why do children, particularly boys, frequently tease animals in cultures around the world? Are we practicing our dominion over the fish of the sea, the fowl of the air, the creatures of the land? Now it is play, but that same intellect will soon abandon all pretext and the game will be for life and death. Years later, after we have lived and suffered and endured the loss of those we loved, animal as well as human, then the thought of what we have visited on helpless creatures makes some of us cringe, for we have come to understand the helplessness in which the animal stands before us.

Despite my ingenuity, however, my favorite tease was simply an act of niggling affection, good for whiling away the idle afternoons, and it was merely to tickle the fur between the toes with a straw or a blade of grass while the dog slept. At first there would be no effect, but soon the toes would begin to twitch, then the foot would pull back in a sleeping reflex, and then, if I persisted, the foot would thrash around to rid itself of the pesky irritation. Finally the dog would raise his head and stare at me with a look of weary patience as if to say "Is everyone having fun?" Usually I would desist, feeling a vague twinge of what I later learned was shame.

One night I tried the toe-tickle with Darwin. He had been shedding his whiskers, and I found a handsome, snow-white specimen lying on the brown floorboards. As I picked it up, memories of my youth came up from the archives, and I could not resist the perverse urge to revisit my transgressions.

Darwin lay on his side, submerged in slumber, and I bent over and began working on his hind feet. At first there was no response. I persisted, of course. Then the left leg twitched, then it jerked, then it kicked out spasmodically. I laughed the thin, whiny laugh of the teaser and continued. Aside from kicking out, Darwin appeared to be unaffected. His eyes remained closed and he seemed to be sleeping, which meant his kicking out was purely reflexive, therefore unconscious.

That was not sufficient. No self-respecting teaser could let it go at that. Any tease worthy of respect had to bring forth a conscious outburst of perplexity, frustration, irritation. Only then would it have succeeded. So I proceeded to Darwin's forepaws and again applied the whisker to the fur between his toes.

As before, the forepaws began to twitch. Sensing success, I increased the tickle rate. The right forepaw pulled in to tuck itself away from the irritation. I turned to the left forepaw. It twitched a few times. I bent over to redouble my efforts. The left forepaw then shot out . . . Followed by the right forepaw . . . Followed by Darwin's entire body, including his mouth, which was opened very wide.

It happened so fast that it took a second for comprehension to set in. When it did, I realized that I was in a predicament. Forepaws — fore*claws* — were grasping my left ankle, which was also clamped firmly in Darwin's jaws. Pinpoint fangs and knife-edged carnassials were pressing into my flesh hard enough to cause pain, but not hard enough to break the skin. Then, to my astonishment, Darwin turned his head slightly, and while keeping his hold, glared at me from the corner of his left eye, ears laid back, pupil dilated. I found myself staring into a black hole of anger which led straight to the bottom of my ancestry, and there it produced a voice.

"How do you like them apples, Pal?"

Now, if you inquire into the nature of intelligence as a biological trait, you soon come to the topic of survival. Like bodily organs — heart, kidney, liver, spleen, and so on — intelligence would

not exist were it not vital to survival upon the face of this glorious blue-green planet. Therefore, the essence of intelligence is
the ability to make the right decision for the circumstances.
Smart is as smart does.

If you follow this reasoning to its logical end, you find
yourself at the proposition that life is inherently intelligent. Nor
does this intelligence require brain, because most living things
do not have brains. Yet they respond appropriately to the world
about them; otherwise they would not exist. The sugar maple, as
good an example as any from the plant kingdom, alters its physiology for the onset of extreme cold, which would otherwise kill
its cells. It sheds its leaves so the snow does not weigh down its
branches and break them off, and it enters the winter prepared
for the forces. Taking this notion to its ultimate conclusion, life
is not merely intelligent: Life *is* intelligence.

As I peered down at my ankle with Darwin attached, I
gazed upon the native intelligence of life and it spoke to me. It
spoke the intuitive language of posture and gesture that lies beneath self-conscious intellect, and its voice was clear and unequivocal. Its message was profound. It comprehended all. The
sense lay in the image, and this is how the logic went:

Darwin wasn't begging me to desist, as a dog would. He
was *telling* me; with his teeth pressing into my flesh he held all
the power. Clearly he was restraining himself and he was doing
so in expectation of my next move. He was negotiating! Having
no time to fabricate a conscious thought, I knew in this flash of
awareness that he had pulled a *coup d'état* and turned the peck
order upside down. Now *he* was dominant.

In the same instantaneous flash, my instincts weighed the
terms that Darwin was offering: My ankle for a few idle laughs

and a trip to the emergency room. A no-brainer, literally. A root
could figure it out.

"Niiiiiicce Daaarwin! Goooooooooood boy!!!!!" I whined. I
bent down ever so slowly, holding out my hand ever so submis-
sively. So very gingerly, so very delicately, I extended my hand to
pet him, watching his eyes for the slightest sign of agitation, and
finally, sensing permission granted, I so, so, very, very gently
touched his head and ran my hand slowly, slowly down his neck
and back, understanding for the first time in my life the thrill of
disabling a bomb.

⁓

We had arrived at a new level in our relationship. I could have
invoked my humanity, put on boots, and continued to tease
him. I could have insisted on dominance and smashed him
down. But that would have ruined a friendship based on the
equality of souls; to force him down would have crippled the
love between me and this small creature.

From this attempt to tease Darwin, I learned that respect is
a tool for dealing with the harsh realities not only of the animal
peck order but of the human hierarchy as well. When you respect another being, you are restraining your self. You curb the
urge to put him down, force him onto a lower rung of the dominance ladder. Showing respect, you defer voluntarily, thus accepting the other as an equal on the same rung. Nor does the
object of your respect have to be a human being; it can be any
being.

I saw, too, that respect exists in two forms or levels. The
first is based on the fear of retribution, as Darwin illustrated
with his fangs, and is therefore forced on others from without.
This is the respect found among animals in their social interac-

tions. It is found widely in human societies in the enforcement of law and among nations in the prominent display of military arms. It is the respect that parents have demanded during five thousand years of civilization in the rearing of children and, until recently, was taught over a parent's knee. The good parent pointed out that this humiliating position could be avoided if the child respected others, respected property, respected the law, respected God, and proceeded down the list of civilized restraints.

The second level of respect is relatively sublime because it comes from within. It is not imposed through the threat of reprisal. You enjoy the act of deferring to others because you have learned that like begets like, and as you treat others with respect, so they respect you. When the day comes that you realize this humble fact — that is the day you become an adult. This too, I learned from Darwin.

9 | HOSPICE CARE

THE WEEKS PASSED. Our bond deepened, grew more subtle. Things were good. One afternoon at the beginning of March, I was sitting at my desk, writing and pondering, when Darwin came down the hall to my office. He walked slowly, tentatively, as if he weren't sure where he wanted to go, and while his movement appeared normal at first glance, something was not right. The sort of thing you don't recognize at the time but notice with shock later, in retrospection. He walked into my office and meowed softly, a sigh of weariness and melancholy. He wandered over to the window behind my chair and crouched to jump onto the sill, where he liked to stretch out and gaze down on the neighbor's garden through half-closed eyes. He tried to jump, but he could not get off the ground and fell back feebly.

He tried again, pawing weakly at the sill, his white booties soft against the surface of the wall. Then he sank slowly to the floor, his footpads leaving eight parallel smudges in long streaks on the flat whiteness of the wall as his paws slid slowly downward.

Such a small, quiet act. The mind's cold blue eye observed,

taking notes. Whatever was happening to Darwin bore no resemblance to the gradual loss of weight and spirit that the feline leukemia virus was supposed to cause. I had prepared myself for that, primed myself to watch for the slow incursions on life, but this display did not match the medical prognosis. Electrified with panic, I ran down the stairs to the storeroom, pulled the transport box from the shelf, strode back up the stairs, and picked Darwin up. He offered himself limply, without spirit, and I placed him gently in the dark interior of the box. I knew then that whatever the matter was, it was serious.

Dr. Mader lifted Darwin up, laid him on the stainless steel examining slab, looked in his mouth.

"His gums are pale — not getting enough oxygen. Let's keep him overnight." With one needle he took a blood sample. With another he gave a shot of tetracycline. Then a young technician picked Darwin up and carried him into the inner sanctum of the medical room. Somewhere within, a cat mewled and a small dog yapped inside their cubicles of stainless steel.

I returned home to a lonely space. Darwin's water bowl sat on the kitchen floor and his food bowl rested clean and shiny on the kitchen counter. Everywhere I looked I saw his limp, listless body in the assistant's arms, the stainless steel door shutting behind them. The cold mind, the rational one, then rose from my body to observe my writhing emotions. Why do these images of Darwin come up, asked the blue intellect? Why does the brain produce clear images of Darwin when he is not here? Why does the brain produce such pain on the absence of a mere cat? The questions were now posted. Subsequent observations, subsequent cogitations might produce the answers.

At nine the next morning the telephone rang. The doctor's voice cut straight to Darwin's diagnosis. His blood had tested positive for a new disease, feline infectious anemia, commonly called by its initials, F.I.A. Feline infectious anemia was caused by a rickettsial bacterium, *Haemobartonella felis,* which attaches itself to red blood cells, inducing the cat's immune system to mistake its own cells for bacterial cells and destroy them, causing acute anemia. That is why Darwin's gums were pale. Feline infectious anemia frequently appears in cats already infected with the feline leukemia virus and it moves fast. If not treated aggressively, it kills in a matter of days. Darwin would have to remain hospitalized until treatment was effective.

The next day I could not stop worrying. All attempts at work produced nothing but lost time, and finally it occurred to me that maybe I could visit Darwin in the hospital. Why did it take so long to think of this obvious notion? Were veterinary patients any different from the human kind, to be shut away in the hospital without visitors until cured and sent home?

Dr. Mader and I had liked each other from the beginning; he respected my knowledge of biology and I respected his experience in medicine, so I called and asked if I could visit Darwin. By all means, he replied. More people ought to visit their pets in the hospital.

⁓

Darwin had a private room on the third level of holding cages, most of which were occupied. In some there was nothing to be seen but a motionless lump curled up along the back wall. In others, a dog, a cat, a rabbit, or some other creature stood behind the bars, waiting with great expectations for its owner.

Darwin lay curled in his litter box, submerged so deeply in

sleep that he appeared comatose. His tail and hindquarters lay in urine that had collected around the edges where the litter had not been scattered. A catheter ran from his penis, and an IV needle was taped to his right foreleg. His breathing was so shallow that I had to peer closely to detect any motion. I opened the door and softly called his name. His ears moved ever so slightly, but his eyes remained closed. Tears welled up behind my eyes and in my throat, and I swallowed hard to force them back. Tears are so naked, so vulnerable. An instinct to suppress them.

I bent over and spoke his name again. His eyes opened just a sliver, but signaled recognition, and I ran my hand over his head and down his neck and back as gently as I could, just skimming the fur. I bent lower so my ear was a few inches from his head and stroked him ever so gently again and again. Then, so quiet that it seemed far, far away, I heard a familiar sound. Darwin was purring. Joy swelled in my chest and squeezed more tears from my eyes because Darwin recognized that I was there. With wet cheeks I stroked and stroked his matted, soiled hair, oblivious to time, reveling in the faint and distant sound of the cat's soul.

I don't know how long I stood there, absorbed in his misery, but this was a working hospital, and life and death went on around us. While I attended Darwin, a large, gray-black cat with bold white markings and light yellow eyes had arrived, following a collision with a car. He was a sweet, gentle creature and he made no attempt to bite or scratch or otherwise resist examination. Abducted to an alien world, he submitted patiently to the pokings and squeezings of a strange man, who said in a flat voice that the cat's left foreleg was broken. The owner then telephoned, and Dr. Mader took the call on a cordless phone. The owner wanted the cat euthanized. She did not want to spend the money for medical care.

Dr. Mader did not cajole or beg her to reconsider. He simply said that the cat would recover fully if treated; he asked if she was sure of her decision; without hesitation the owner said yes. Thus was the cat's fate decided. Dr. Mader then turned to one of his assistants and said matter-of-factly that the owner wanted the cat put down. The assistant, a thin, sensitive young man, nodded and set about preparing the equipment. The other assistant, a quiet young woman, held the cat in her lap while the young man shaved a small area on the inside of its right foreleg. Both assistants bowed their heads in a silent, dignified gesture to this lovely creature who was about to die, as if to say "We love you. There is no other way." Then the young man administered a hypodermic injection. The cat lay there, calm, as the young woman held him in her arms and stroked his head and back. Gradually he grew drowsy. His muscles relaxed, almost imperceptibly. His eyelids began to droop, then his head sank slowly onto the young woman's lap and lay there motionless, with his eyes half open . . . and lay, and lay . . . until there came an instant when you somehow knew his soul had slipped silently, invisibly, into eternity. No raging against the dying of the light. A simple passing into the light.

⟶

This big, beautiful, vigorous, healthy cat, and the owner had elected to end its life because treatment cost money, and here was my Darwin, his body curled in mortal combat against his own immune system. Tears had irritated my nose and made breathing difficult, and the hopelessness of it all pressed down with an aching, suffocating weight. The blue intellect hovered above, pointing out that health is like wealth: you cannot take it with you. But unlike wealth, you cannot give it away. I would have paid any price to buy the health of that poor, unwanted cat and give it to my friend.

Breathing deeply, I managed to reinstate my composure and regain sight through the liquid blur. Then I noticed a large red tag on Darwin's cage. "FeLV POSITIVE," a warning to regard him as highly infectious, a carrier of the plague, a pariah.

When I called the hospital the next day, Dr. Mader informed me that Darwin's spirits had lifted noticeably after my visit. He was still in critical condition, however, and I visited him again that afternoon. This time he recognized me as soon as I spoke and leaned into my hand, lingering and savoring my touch.

The visits became a daily ritual. Each afternoon I spent an hour by Darwin's side, stroking, talking, but mostly spending the time quietly, giving thanks, relishing the time we had together. Despite his illness, he still lived, and for the first time in my life I began to comprehend the miracle of each living moment, to *feel* the complexity of a living organism. My knowledge of biology slowly emerged from years past and I began to remember the incredible physiology of the blood, the nervous system, the kidneys, the liver, the spleen, the heart. I savored the warmth from Darwin's body and recalled the basics of temperature regulation and water balance, and then, of course, the immune system and physiological defense. Words could not do justice. What was Darwin but a universe in a skin?

Recovery proceeded slowly. Darwin's temperature remained high. The IV remained in his foreleg, bleeding tetracycline directly into the bloodstream to combat the *Haemo-*

bartonella microbe. As one day led to another and another, the costs added up, although I didn't know what they were and made a point not to ask. I was going to give Darwin the best care possible — that was not negotiable with my rational self — and I would simply have to deal with the cost later, as a separate crisis.

The emotional costs were also accumulating, probably more for Darwin than for me, huddled as he was in his stainless steel sanctuary. Strange hands opened the door several time each day, often to slip needles under his skin for various injections, now and then to draw blood from the carotid artery in his neck. The intravenous needle was repositioned from time to time and the catheter removed for relief. Alien animals occupied the cages above him and on both sides, emitting alien sounds and alien odors, and occasionally shuffling about in search of a comfortable position. What does a creature feel on entering the hospital, knowing only blind, instinctive terror and incapable of conceiving the reasons for his internment? The enormity cannot be imagined. That it saves his life means nothing when he cannot conceive of death.

As for me, I wanted to deny the undeniable. The lower, more primitive depths of the mind cannot face the notion of death because the spirit of DNA rises up from the neurons of the brain, coalescing into the genie of the human mind. "Thou Shalt Survive" is the First Commandment of DNA, and each cell obeys it blindly and mechanically, as does the genie. Yet that same mind holds the capacity to comprehend death. It knows that we will die someday. And suddenly we find ourselves staring into the existential dilemma — that nuclear furnace of paradox — where the deepest essence of life, the will to survive, comes face to face with the truth of reality, and the conflict is too intense to face. It is the curse of biological intellect.

The upshot? Why gods, of course, gods created in our image and projected back upon the world. Very large, very powerful mamas and papas who understand our needs and look out for our interests, assuring us that the soul lives beyond the body, and giving us dominion over the beasts and fish and fowl. These self-serving projections comfort us as we walk through the valley of the shadow of death and hold us back as we peer over the edge into the abyss.

The gods were small consolation, however, since those of the West have little time for the plight of animals. The only answer was hope, which arises from ashes. Hope is closely allied with augury, an attempt to project the future, and medical science is rife with augury in the guise of data and statistical projections. Body temperature is an excellent example of how this works.

The normal cat maintains a body temperature between 100.4°F and 102.5°F, and although body heat, like breathing, is something we take for granted, it requires a miraculously strong, healthy physiology in order to maintain a steady level. A normal body temperature is therefore an encouraging omen, and each day my first act on arriving at the hospital was to check Darwin's chart for the latest temperature reading.

On the first day it read 103°F, and even though he was nearly comatose, this was essentially a normal reading, a good omen indeed. On the second day it jumped to 105°F, but that was easy to explain. A fever was to be expected in the seriously ill. On the third day the temperature dropped to 102°F, and I walked several inches off the ground on pure euphoria. To hell with the calculating side of the mind, which urged caution in its usual sour way. On the fourth day the calculating side proved right as Darwin's temperature jumped to 104°F and my spirits dropped. No pagan ever stood by his fire, examining guts in the sand, with more trepidation than I stood before Darwin's data attempting to twist the prognosis in our favor.

The same emotions heaved and pitched according to Darwin's weight. Overall he was gaining weight, and hope clutched this as a major sign of improvement. It helped me minimize the forlorn image of Darwin harnessed to medical technology, reminding me of Edward Abbey's immortal description of patients in the terminal ward with "rubber tubes stuck up . . . nose, prick, asshole, with blood transfusions and intravenous feeding . . . the whole nasty routine to which most dying men, in our time, are condemned."

So it went, temperature up one day, normal the next, weight returning to its original point, then dropping below, with hope pulling each observation, each reading, toward a positive interpretation. Finally, after ten days, Dr. Mader released Darwin to my care, accompanied by a bill for several hundred dollars, which still represented a significant discount. Had the animal hospital charged for each cotton swab, each cotton ball, each grain of cat litter, in the spirit of a human hospital, the bill would have been considerably higher.

And it would still have been worth every dollar. Darwin remained a very sick cat, he would require medication for at least another ten days, *but he was alive,* and surely he would improve. As I carried him from the building, I thought of that handsome cat whose broken leg had brokered his death, and I found myself wincing. I winced at the innocent helplessness of animals, huddled at the feet of humanity . . . the absolute ending of life . . . the finality . . . and then I realized that because I had come to love this small creature, whatever happened to him happened to me.

On arriving home, I opened the carrier and Darwin walked unsteadily into the room. He looked around as if home were strange territory and stood there, wobbling, feeble, bewildered. His coat was matted and dull, his white belly stained yellow. I walked quickly to the kitchen and opened a can of food, knowing that nothing would catch his attention like the sweet song of the opener. Darwin remained where he stood, staring blankly. I dished several scoops of tuna into his bowl and set it before him. He looked down for a moment, then bent down to investigate. He sniffed indifferently, pulled back, sat up straight.

I pushed the food toward him. He pulled back, turned slowly, walked away. I could not call up his appetite, which stripped me of the power of manipulation and reduced me to a failed human being, feeling helpless and bleak.

Darwin understood none of this. He walked dejectedly to my leather recliner, my one decent piece of furniture, jumped onto it, digging his claws into the surface, curled up, and went to sleep. I went to the linen cabinet, folded a thick, soft bath towel, carefully picked him up, arranged a bed on the leather chair, and laid him on it.

He did not sleep with me that night. Perhaps he simply preferred the chair to my bed, or perhaps, in his sickened state, my tossing and turning would have bothered him. I could not tell. I missed his warm body pressed against my side, and yet, to know he lay a few feet away in the darkness filled me with sweetly sorrowed gratitude as I drifted off to sleep.

The next day he pecked at his food, eating as if chewing were a chore. I called Dr. Mader, who recommended baby food: ham, lamb. This required only licking, and I fed Darwin from a spoon, like an infant. He consumed perhaps two-thirds of a jar, not nearly enough to maintain his body weight, but he chose to sleep on the bed with me that second night, his warm lump softening the bleak realities with a sign of recovery.

On the third day it was time for a checkup. I lugged the carrier into the living room, set it on the floor, swung open the door, and forced Darwin in. He huddled at the far end of the cage, pupils dilated to the edge of the iris, and stared at specters of the hospital with black eyes.

Dr. Mader pulled him from the darkness of the carrier, set him on the scales, and wrote down 12.6 pounds. He inserted a rectal thermometer and wrote down 105.5°F. My rational mind watched without feeling as these figures appeared on paper — temperature was high, weight had dropped by more than a pound.

Throughout the procedure, Dr. Mader showed no emotion. In the praxis of medicine, to indulge feelings is to waste time. With the emotionless calculations of medical cause and effect he explained that Darwin did not need intravenous feeding and he didn't need a catheter and he didn't need incarceration. He needed only to take an antibiotic called Tribrissen. It was more powerful than tetracycline, came as a pill, and would have to be administered four times daily.

Now, finally, I had to face the challenge of forcing pills down the armed throat of a cat. It was a task I had been dreading since I first read the handbook of veterinary medicine.

When you open a cat's mouth, what you see are thirty polished, razor-sharp teeth. There are fourteen molars and premolars that form blades for slicing cartilage and gristle and bone, which is why they are called carnassials. There are four fangs at the corners, which function as picks for stabbing prey and keeping it pinned. There are twelve tiny incisors arrayed across the front, six in the upper jaw and six in the lower, which have evolved to destroy fleas; they function as a comb. Everything about this device is designed for carnage, and as you hold those jaws apart, the racial memories and primal fears of our simian forebears — the ones who survived, that is — waft up on the cat's fetid breath. This cat, too, had ancestors, and many of them were very, very large.

Clearly, to administer Darwin's pills I would need some instruction, so I turned to the bible of veterinary medicine. The first step, according to the good book, was to recruit an able-bodied assistant and, at the first signs of resistance, to roll the cat up in a towel like a burrito. I decided to call Robyn, my angel of mercy, who rescued animals as a daily matter of course and was fearless in nursing them. She halted traffic on four-lane streets to protect bewildered dogs and cats, stopped on freeways to pick them up, saved them from the pound, placed them in homes. She had a feeling for small creatures, and small creatures for her, that I had always trivialized as saccharine sentimentality, but desperation had completely reversed my judgment and now I found myself in supplication before her. Robyn would show me how to navigate Darwin's gauntlet. At a glance from Robyn, small animals rolled on their backs, opened wide their mouths, and *begged* for pills.

As I expected, Darwin submitted to her graces without the slightest hesitation, and before I could blink she had opened his mouth and poked the pill down his gullet in a motion so swift, so sure and flowing, that I could not revisualize it and certainly not repeat it.

Since we could not give another pill until the next dose fell due, Robyn coached me, demonstrating how to place the left hand over Darwin's head, how to place the right index finger on his lower lip and slowly pry open his mouth; she showed me how to grip his head in my left hand and press the index finger and thumb from opposite sides behind his molars, wedging his jaws apart. And with her right hand, she performed the *coup de grâce* — placing the pill as far back as possible and poking it down the gullet. Finally, she demonstrated the endgame, gently holding Darwin's mouth shut and stroking his throat to activate a swallow reflex and conduct the pill to its gastrointestinal destiny.

The next morning the time came to put theory into practice. I set Darwin between my knees, so he could neither pull back nor bolt to either side, and set the pill next to him on the floor. Very gently I gripped his head with my left hand as Robyn had showed me, pried his jaws open until his mouth was gaping wide and I was peering into an abyss of wet, pink, glistening flesh studded with teeth, teeth, teeth.

Then Darwin began to panic. He tried to twist his head from my grasp. Instinctively my eyes zoomed into a macro view of the writhing tongue and the huge, straining masseter muscles that made the jaws clamp down. All subsequent memories are blurred, but I vaguely recall my index finger and thumb pole-vaulting over the middle finger and shoving the pill into the abyss. The tongue heaved and the pill disappeared. I jerked my hand away and fell back, thinking a prayer. Darwin, however, just sat there without moving, ears laid back, and slowly licked his lips.

Right from the beginning he seemed to regard these proceedings with a certain detached nonchalance and took pills easily. He never lost control, never reverted to the lethal stabbing and slashing and gouging machine that a mature cat can, under stress, become. Whatever the reason, I thanked the fates for small mercies and apologized to any I might have overlooked.

⁓

The next morning I went to the kitchen and tried again to interest Darwin in food. I pulled open the utensil drawer, removed the can opener, clamped it onto a can of tuna, and cranked it around the perimeter with an exaggerated commentary on the odor of cat food. "Oh, boy!!! Your favorite treat," and so on. Darwin, lying on the bed, opened his eyes. A few seconds later, his ears swiveled sluggishly in my direction. But he made no attempt to move, and my heart sank. A few weeks earlier he would have read my movements even before I entered the kitchen, and come trotting to my side. He would have sat down at my feet and stared up at me, focusing a beam of pure, animal hunger on the god of food.

The cool blue mind understood this. It knew with quiet desperation that Darwin could not be expected to eat.

"Oh, *boy!!!*" I implored. "Your favorite *food!!!* Here it *is...*" but the words sounded empty, phony, and trailed off into elliptical dots.

His eyes closed, and slowly he laid his head on his forepaws.

I held the food under his nose. He turned away. I followed with the bowl and again placed the food beneath his nose. He turned back the other way. I was about to continue my efforts when the thought occurred that maybe he really didn't want to eat and I should respect his refusal. Reluctantly, I controlled myself and took the food back to the kitchen.

Over the next two days, Darwin showed little interest in his food, pecking listlessly while I stood by and pleaded that he eat. Clearly, his intake would not sustain him in the long run, and his weight would soon begin to drop. On the third day, I took him back to the hospital for another checkup. Three days later, still another checkup, this time with another blood test, which came back negative for *Haemobartonella,* meaning that the results were positive. Dr. Mader concluded that Darwin was continuing to recover.

In the following weeks Darwin's appetite gradually increased and slowly his health improved. It was now well into spring, and the average temperature was rising by the day. He began to resume his territorial patrols. In the afternoons he took to his hammock in the billowing folds of the Volkswagen cover, his plump little form merging into the sensual, breathing shapes of the wind-inflated cloth. Given such progress it was easy to abandon oneself to hope.

Then one day at the end of April Darwin began to droop again. He came down with diarrhea. He vomited. A grim, quiet hysteria rose up from the blackness beneath my mind, for the most dreaded of all chores was surely at hand. I would have to put Darwin down.

No. No. Not without a fight. I would take him to the hospital. But if he was dying I wanted to hold him in my arms, wanted to savor the warmth of his body one last time. I needed someone to drive us.

Who to ask on a weekday afternoon, when decent people are at work? I ran down my list of friends again and again. Then I remembered Suzanne. I had met her only once, after she had written a passionate letter to the local newspaper advocating a ban on the hunting of mountain lions. It had so affected me that I looked her up in the white pages and telephoned her the next day.

She was a true lover of animals, cats in particular, and as soon as we met, I sensed she was a person of integrity, someone to cultivate as a friend (not, unfortunately, as a lover; she was living with her fiancé), someone I could call upon for help with a dying companion. She was a student and she happened to be home.

Suzanne arrived within ten minutes and suggested we take her car, an old rusting Volkswagen with springs coiling up from the seats and rusty metal floorboards where carpeting once lay.

Clutching Darwin, I sat down on steel coils as the air-cooled engine roared into life, choking and coughing, with a backfire thrown in for celebration. Darwin responded with a stream of urine, which, because it was hidden beneath his body, came to my attention as a warm rivulet running down my leg and a pool of yellow liquid spreading across the rusty metal floor.

"Oh, God, I'm sorry, Suzanne. We should have taken my car. I'm sorry."

"Please," said Suzanne. "That's what this car is for."

We found parking directly in front of the hospital, and with wet blotches covering my lap like camouflage, with yellow liquid dripping from my cuffs, I carried Darwin in one continuous hug into the office. Dr. Mader soon emerged from his medical sanctum. I managed to choke out my grief.

"I think . . . it's time . . ."

He took Darwin from my arms, weighed him, laid him on the stainless steel table, looked in his mouth, checked the whites of his eyes, palpated his abdomen, took his temperature.

"Are you sure you want to do this?" he asked, with a concerned frown. "I think we can fight it."

I had spent the morning in premature grief, facing the worst, and it took a few seconds to register the shock of possibility. Of course I wanted to fight the dying of the light; I'd do anything to save the creature I loved. A needle penetrated Darwin's skin to inject antibiotics, another needle entered a vein to draw blood.

The next day the results showed no evidence of *Haemobartonella felis*, only a strong positive for the feline leukemia virus. On the second day Darwin reentered the hospital with vomiting and diarrhea, diagnosed with "nonspecific enteritis." This, apparently, was the path we would travel from now on: driving to and from the hospital as the leukemia virus gnawed slowly away on Darwin's life and forced him time and again into these episodes.

That night, with Darwin in the hospital, the rooms seemed so empty, the walls so bare, and the truth pressed in from all sides with a black, suffocating weight like deep water. I realized then that sooner or later Darwin would have to be euthanized for the sake of mercy. It was the truth I dreaded above all others.

10 NIGHT WALK

SLEEP WAS OUT of the question. I was reaching the limits of emotional endurance and could no longer face the acrophobic peaks and pathetic depths of hope and reason. The time had come for a summit meeting of the minds. The heart would have to be reconciled with the head, the feelings and urges of the limbic brain reconciled with the calculations of the rational cortex. It was, in other words, time to set standards by which to make that final decision.

It had rained off and on during the day, but the rain had fallen off about a half hour earlier, and I decided to take a walk. There is something in the striding of the legs, the pumping of the chest, the beating of the heart, the buildings and trees passing before the eyes, the mood of the light, the temperature, the humidity — something in the rhythm and the physical exertion of simple walking that clears the mind and, at its most productive, makes for revelations. Zipping my jacket and raising my hood, I walked down the stairs and headed into the night, toward the sea.

The air was cold and damp, with sprays of light rain now

and then tingling against the skin. The blue light of the mer-cury-vapor streetlights glinted off the wet asphalt and glared from the sidewalk puddles, and yet, despite the reflections, the streets felt much blacker than usual. It was as if the rain, which lay in a film of molecules over the world, was absorbing more light than it reflected back. Rainy nights always seem so dark, dry nights so much lighter.

For ten years I had lived in this conservative, unpreten-tious neighborhood of Craftsman cottages known as Belmont Heights, a half mile from the long, wide beach that gave the city its name, and I walked with memories playing simultaneously with the present reality. At the corner of Loma and Third I came to Temple Beth Israel, a Jewish atoll in a sea of Christian bunga-lows. I thought of the morning several months before when the groundskeeper appeared at my door and handed me Darwin's address collar, which he had found on the lawn. A fight had probably occurred, the collar had been lost in the scuffle, and the groundskeeper had taken the time to return it.

I continued toward the sea and turned right on Broadway. I passed the pet shop with its bins of puppies displayed in the window, produced, probably, in some midwestern puppy mill, pedigreed with genetic flaws, and cunningly placed to manipu-late the heartstrings of children and kindly people. I passed a pub, now having some Irish name, but seeing the faded sign KARL'S LITTLE BAVARIA over the door in honor of its original owner and founder, Karl Kohlbrecher.

Herr Kohlbrecher had survived World War II as a profes-sional wrestler and boxer in the personal employ of one Adolph Hitler, who retained him to perform for small, private audiences that appreciated his skill in braining his opponents. Following the war, he immigrated to Long Beach and opened this dingy little window on the Reich one block south of Temple Beth Israel.

I had entered it once — Karl was long dead of a heart attack behind his beloved bar — to buy potato salad, for it was widely rumored to be the best in town. But the potato salad was only a ruse. I entered Karl's Little Bavaria with the German half of my ancestry peering through slots between fingers, dreading what I might see, or feel, yet drawn like a moth to black light. I vaguely recalled pictures on the wall of Karl with members of der führer's peers, although I may have been fabricating memories, and though I could not remember which peers they were, I did remember how dreary and dark and reeking of stale beer the place was, an apt monument for the men it conjured up.

I walked on and Walt Whitman came to mind. "I am large. I contain multitudes," exulted the poet of the people, praising the human condition in its vast, incomprehensible sweep. Did he write that immortal line before or after his service as a wartime nurse, tending the maimed and dying on the slaughter fields of the Civil War? What song would Old Walt have sung had he known the miracles of modern biology? The line kept repeating: "I am large, I contain . . ." I thought of intestinal bacteria, the true guardians of the natural world. Yes, indeed, I contain multitudes, and I walked along, contemplating the matter of one small human clutching the life of one small cat, wondering wide-eyed at the insignificance of individual life. For hatred, for love — what difference does the taking of life make to the molecules, of which we all consist?

I could not reply for the molecules, but to Darwin and me, the answer made all the difference in the world. To snuff the life of this small companion cut against every fiber of my being, a response I would not have believed a half year before, and this feeling was intensified by the nature of the love between human and animal. To take the life of my friend, no matter how sound the reason or merciful the intent, would violate the innocence of our bond and desecrate the intimacy with which we had shared our existence. Even to contemplate the act knotted my stomach.

However, despite my aversion, if Darwin's suffering became undeniable, his misery overwhelming, no decent human being could allow it to continue, and I would do what had to be done. That, too, was a visceral response. But when the time came, I wanted mercy to meet every possible standard of dignity and compassion.

On I went, splashing through puddles, shattering reflections, and by and by I began to sort out the triage of mercy. The essence of a friendship, and particularly of a loving friendship, was free choice. To end Darwin's life without his consent would be to play god. And yet . . . and yet . . . what choice did one have but to make autocratic decisions in one's relationship with a cat? Was it even *possible* to avoid the role of god? Whatever we do or do *not* do, the animal has no choice in the matter, making us gods if we do and gods if we don't. Malicious god? Benign devil? That, apparently, was the choice.

I walked along, spiraling inward with dismal meditations on dying. If we all proceed inexorably and ineluctably toward death, is not living a lifelong process of dying?

This was not the time, however, to submerge myself in metaphysics. A small cloud of spray swept past, breaking the spell, and I decided on some simple, practical criteria to help in making the decision, the heart of which was to recognize the right time, to see the proper moment to free the spirit from its misery. For our purposes, it seemed that dying would be the sudden decline, accompanied by chronic misery and pain that outweighed the desire to go on. Was I extending life, or prolonging death? That was the question I would have to ask, and Darwin answer.

I turned left on Redondo Avenue and headed once more toward the sea. As a god of mercy, how would I apply my art? How would I know when the time had come? At the corner of Redondo and First the epiphany came through. If I had to play god, I would be a selfless god! *It was Darwin who was dying, Darwin who suffered. My own feelings meant nothing. My purpose was to nurture and protect him, not to indulge self-pity.* I would examine my thoughts for signs of self-indulgence and, when discovered, deny them without appeal. I would focus every scintilla of sensitivity on Darwin's actions, his postures, his cries, on the subtlest nuance of his life. Then, having transcended my self, I would see the world through his eyes and I would know his feelings. My decision would be his decision, more or less. I would help him so long as he wished to fight for life, but in the end it was Darwin who would have to say "Enough."

With the cold spray stinging my face and warm tears tightening my throat, I came to the cliffs at Bluff Park and stopped at the guard rail to simply exist in the raw, black air. I stared out over Long Beach Harbor, over the artificial oil islands with their wind-burned palms and grotesquely lighted condo façades disguising the drilling towers and pumps, over the *Queen Mary* and the immense geodesic dome of the *Spruce Goose*, over the

leviathan tankers and cargo ship waiting to disgorge their contents at the docks to feed the titanic gantry cranes and the re-
finery terminals, over the breakwater, over the absurd into the
sublime — and drifted off, into the eternal blackness on the sea
at night.

Arriving home, I climbed the stairs and opened the door, a god
of the intellect with my conclusions rolled up like a scroll into a
neat, tidy covenant. Without a clue as to where my philosophies
were about to take me, I sat down and began to grieve. Grief, it
appeared, began not with death but with the recognition of its
imminence and made itself manifest in a deep, throbbing blend
of pleasure and pain. From this point on, Darwin and I would
become intimate companions with this sweet sorrow.

What I found in the months to come was that in focusing
all my powers on Darwin's state, I became sensitized to his every
cry, every gesture, every marking, every hair misplaced, every
whisker shed, every smudge on his perianal fur, every crust of
dried tears, every drop from a runny nose, every utterance, even
the softest suspiration.

During the day I doted on him, glancing in his direction
countless times, changing his water with each drink, removing
his droppings and peepee potatoes with each visit to the litter
box, coming to his side at each meow, racking my brains to an-
ticipate what else he might possibly need. Wherever he chose to
sleep, I would make a bed of towels on the spot. With each listless movement I would check his gums for the pale evidence of
feline infectious anemia. I even slept more lightly, somehow
alert, and if Darwin cried out in the night, his voice sliced like a
razor across an eye. Scalp crawling, half awake, I would jump

from bed, run to his side, hoping — praying — that I could help him, and when done, feeling that serving his needs was not a chore but a privilege, and a glowing sense of gratitude came over me. I refused invitations, restricted my travels, rearranged my sleep to administer medication, spent hours preparing his food in special ways.

In serving Darwin I denied my own immediate desires and in this denial found an unexpected sense of self-worth. There was a power, it seemed, in curbing the animal urge, from controlling the primal appetite, by means of rational determination. Monks and priests, leading lives of religious self-denial, have always known this. In a similar way I discovered the spiritual intimacy of terminal care, which allows no modesty, no shame, no conceits: messes would be made; messes would simply be cleaned up without further concern. I trudged on, bearing the weight of impending death. To know that this small, doomed creature would have his every need met, his every pain caressed, transcended hopelessness, and often I would find myself swallowing tears of sorrow and joy, for this was truly the nectar of love. That is what happens to gods who encounter cats.

Inevitably the cycle of daily life turned on feeding. When Darwin ate, the sun came forth. When his appetite waned, the day turned black, the virus was surely encroaching, the end was drawing near, and the grief throbbed with its heavy, pressing ache.

And so I badgered him with culinary temptations. I cooked salmon and chicken and steak (although much of the time I succumbed to my own appetite and simply made larger amounts, with enough to share). Sometimes Darwin would gulp his food with smacking lips and sucking gusto. With increasing frequency, however, he would eat more slowly, with less joy, then turn away and leave food behind. He had never done that when his health was good, and the memory of better times always made me sad.

A pattern emerged in which his appetite rose and fell in a seven- to ten-day cycle, almost certainly reflecting the viral encroachments. One day, as one of these cycles approached its nadir, with appetite nearly gone, I realized that he had not moved for hours from his spot beneath my desk, and on reaching down to stroke him, I found that he lay in a pool of urine with several large stools. When I scooped him up, he lay limply in my arms.

By now I was learning to temper my panic at moments like these, and instead of driving straightaway to the hospital, I called Dr. Mader, who recommended that we wait a day or two to see how Darwin fared on his own. He was taking antibiotics, which would prevent the feline infectious anemia from flaring up, and no treatment would suppress the virus, so there was not much to be done.

After a sponge bath, Darwin squeezed beneath the couch, implying that food was out of the question, and curled up to endure. For two days he lived in a state of deep depression, listless and inert; then, on the third day, he got up, walked to the door and announced with a loud, paint-peeling meow that he wanted out. After all, if he did not patrol his borders, no one else would. He walked down the stairs, tail up, as if nothing had happened, and for six or seven days he appeared to enjoy normal health.

A few days later, Dr. Mader called and asked if I'd be interested in exposing Darwin to an experimental treatment. The medication was called immunoglobulin, a natural substance that was supposed to stimulate the immune system. There were no guarantees, but preliminary tests had shown promise in fighting the feline leukemia virus. The program would involve intravenous injections every three or four days and would require six treatments or more.

I looked at Darwin lying in his corner, his bold markings and rich orange-and-white coat still stunningly beautiful, and saw him looking back at me. His head was resting on his paws, but his eyes were turned up, looking out from under his brow. I tried to focus my spirit, tried to enter his skull, get into his mind, feel what he was feeling. I felt so . . . spiritual — surely our minds would connect. But his eyes stared back at me without any sign of emotion or thought, just those large, black, slitted portals leading back to that place where evolution began. All was blank, all was isolation, nothing was said. The realization slowly sank in that any dialogue we had would have to be manufactured by me.

Very well. He still has a lot of living left. He might be terrified of the hospital because he cannot understand the reason behind the torture. He cannot think these things through, but I, with my power of durable human attorney, can. What if the treatments work? We'll do it.

The next day we began. Transport cage, car, hospital. The usual terror. Another needle in the leg. Several hours to infuse the liquid. Transport cage, car, home.

I could see no improvement in Darwin's bearing, mood, or overall appearance over the next few days, and God knows, I wanted to. The blue mind, however, would not fall for sentiment. Four days later, after the second treatment, right on the seven- to ten-day cycle, Darwin retreated beneath the couch and curled up with his tail wrapped around his forepaws and nose and entered another depression. His temperature remained normal, but his heartbeat was slightly elevated. Perhaps the time had come, said Dr. Mader, to monitor his heart with an electro-cardiograph. This would cost another $140, but I had proceeded too far into the land of medical illusion to stop now. Little patches of his fur were shaved away in each armpit and each knee, electrodes were attached to the bare skin, he was laid on his right side on a padded table and underwent the procedure. He was too weak to protest.

The electrocardiogram, however, revealed nothing significant. It provided $140 worth of knowledge that left us precisely where we started before coming to the hospital.

And so the immunoglobulin treatment continued. Every three or four days I took Darwin to the hospital in *rigor terroris*. On the seventh visit, Dr. Mader announced that the cephalic vein of Darwin's left leg had become so scarred from repeated punctures that we would have to use the right leg. This would be the last injection. It was plain for all to see that Darwin was not improving, and there was no point in continuing the tyranny of medical expectation with its diabolical inflictions of hope.

As if to underscore the viral encroachment, Darwin contracted ear mites. If this seems incongruous, the fact is, as the individual grows weaker, it succumbs more readily to metazoan parasites like mites and fleas. Perhaps there are factors in the blood of a healthy animal that repel these invaders; perhaps a healthy animal can defend itself by grooming; perhaps there are things in heaven and hell undreamed of in our medical-physiological philosophies. Whatever the reason, the mites were the first of what was to be a continuing succession of afflictions mediated by the feline leukemia virus.

The treatment required me to clean the sticky brown mite

secretion from the ridges and recesses of Darwin's ears, then apply medication with a cotton swab. This had a happy consequence, however, because it imposed an unexpected intimacy. In order to do a thorough job, I had to hold Darwin's head and gently tilt it this way and that to reach the hidden crannies of the outer ear, and it seemed as if my fingers, my skin, my muscles, and even my joints were sensing some sort of spirit in the fur, the skin, the muscles, and joints of my fur-bearing friend. The connection was deeper than intellect, the limbic workings of my primal brain fusing directly with the animal mind of our kin.

11 | HOME INVASION

JUNE HAD NOW ARRIVED, the days were reaching their greatest length, and the weather, reflecting the infusions of ever more energy from the sun, grew warmer and warmer. One day I sat in my second-story office, gazing at the meticulously coifed yard of my neighbor, when a large, dark tabby appeared from under the far fence and walked cautiously toward my flat. The neighbor had just watered the lawn, and large puddles covered much of it. Clearly the cat did not want to get his feet wet, so he groped his way, placing one foot gingerly ahead of the other, testing, fastidiously shaking his paws whenever touching the water.

His feet were smaller than one would expect for a cat his size, and this enhanced the impression of refined, almost decadent sensitivities. Whatever the reason, he struck a subliminal resonance in me, and I liked his cautious approach. I understood his reluctance to tread through swamps. But although it was clearly a case of attraction at first glance, I didn't think past the moment, presuming it to be a fleeting fancy. Darwin was more than enough responsibility.

Over the next few weeks I saw the cat on odd occasions and assumed he was a stray. A large apartment building stood on the other side of my neighbor's backyard, and every year when the spring term ended at the local college and university campuses, students would leave for the summer and abandon their cats, presuming that some neighbor would take them in. Students, however, had no exclusive claim to this behavior, since ordinary tenants also abandoned their cats for the same convenient reasoning, and new strays showed up continually, some surviving on the good graces of neighbors who fed them, others simply fading into some land of the never-never that no one knew and no one cared to know.

This was a handsome cat with black spots against a gray-brown background, the ancestral markings and coloration of the European wild cat from which the house cat is descended. In the world of cat fancy, these markings and coloration are known as "standard tabby." His face bore the classic feline mask, with one fine line running from the outside corner of the eye and one from beneath the cheek, meeting at a point below the ear. Four fine lines — the same as Darwin's markings — started at the eyebrows and swept back, coalescing in a black cap atop his head. Five heavier lines trailed back from this cap and ran to his shoulders, where they came together in a wide black band that continued along his spine to the tip of his tail. Black stripes hung down from the band like ribs along his sides, but the stripes were broken at regular intervals, creating a pattern of spots arranged in columns. Along the tail, the band connected a series of black rings similar to the markings of a raccoon.

It was the spotted sides, however, that set this cat apart from the standard standard tabby, because they vaguely suggested the markings of a fish in the genus *Scomber,* known in common speech as the mackerel. He was a member of that small, exalted circle known as the mackerel tabby, and what a fortuitous fate for a cat: the mackerel is a small tuna; and for tuna the cat will risk life, limb, and the pursuit of sleep. By and by these facts would converge on his destiny.

Late one afternoon I saw this cat lying on his side in the dusty, bone-dry yard of the neighbor on the north side — the one who owned the bougainvillea bush where Darwin and I met. At this time of day the sunlight highlights the red in things, and it struck me how closely his belly fur matched Darwin's overall coloration. For all I knew, the two could be siblings — stranger things happen in the society of cats — but I couldn't help wondering, in a twinge of bemusement, if there was some cosmo-comic significance here.

Meanwhile, Darwin was staying outside more and more of the time. In the evenings I sometimes had to pick him up and carry him indoors, virtually dragging him away from his territorial patrols and the perpetual standoffs/skirmishes along the borders. I never knew which of the strays was laying siege to Darwin's estate, but I knew they lurked everywhere and I lived in constant concern for Darwin's health.

I did not shut him indoors at all times, because I knew how persuasive he could be when he wanted in or out, and I had no other place to live and work. Even were it possible to keep him indoors, I would not have done so, because he loved the slings and arrows of the feline military existence. Combat gave meaning to his life. Danger was what he lived to face. Putting his wishes ahead of my own, I clenched my jaw, bit my tongue, and swallowed the anxieties, the fears, the apprehensions, along with the urges to manipulate and control that made me human. I forced myself to endure this torment of possibly losing him, because I understood from biology that life has evolved to endure pain, suffering, harassment, aggression, disease. I accepted the dangers of the world as not only the price for being alive but the reason for living. So I let him in and I let him out as he wished (within reason and with the exception of the nighttime curfew) because the price of equality is autonomy.

One evening Darwin came in for dinner with a nasty red gash running down his muzzle in scarlet contrast with his white fur. On the one hand, he must have been feeling good enough to fight; on the other, such wounds were open portals to any number of microbes, and this red-on-white cut raised the specter of infection. Given his compromised immune system, another infection was the last thing Darwin, or I, needed.

The next morning, after releasing Darwin to the world, I happened to look out the window, and what did I see but the new cat walking slowly, menacingly to the left. From the left who should be walking slowly, menacingly to the right but Darwin. They stopped about two feet apart and faced off. A low moan started soft and then rose from Darwin's throat into a big, cutting yowl, a sword drawn slowly from its scabbard. Tails slashed back and forth in random fits, and fur stood on end like bristles on a brush.

Suddenly I knew who had slashed Darwin's face, and he was about to do it again as the two squared off. But Darwin was in no shape to fight this young, fit, street-tough stray; he was old, probably older than I thought, and sick and fat and hopelessly out of condition. He just didn't know it.

My beloved Darwin. Fury rushed up from the evolutionary

depths of my brain, and I ran to the storage space behind the Murphy bed to rifle through my junk, groping for the Pocket Rocket slingshot with magnum rubber bands. This was not a toy for children; it was a serious weapon that could be used to hunt small game. I grabbed a handful of large marbles for ammunition and bounded down the stairs to the battleground.

Neither cat noticed me. Each filled the other's mind and obliterated all other perceptions. Crouching low, I sneaked behind the intruder so that I was looking at his back and facing Darwin. With a predator's mind, clear and blank of compassion, moral values, inhibition of any sort, I placed a marble in the slingshot's leather pouch and drew a bead on the strange cat's back. I aimed for a spinal shot about halfway down, in the middle of the large black band, paused, and released. The marble flew straight and almost true, but just a fraction off center, thudding into the back about an inch to the right of the spinal column.

The cat sprang into the air and bolted directly away, flinging clods of grass into the air. It careened around the far corner of the house, the place where Darwin and I had met, and disappeared from my view, crying as it went. The cries came drifting back, pain fading into the distance.

The sound stayed in my brain, repeating itself again and again. I would never have reacted this way before joining life with Darwin. As a hunter, the thrill of stalking, of swinging the muzzle onto the arc of flight, of finding the precise, magic instant to jerk the trigger, the thunder of the round and the blow to the shoulder, the feathers exploding from the target, the crumpling in midair, the delirious charge to secure the kill — such sensations had always blocked all thought and reflection.

Now I could not escape that haunting cry of pain. I had almost killed this innocent creature, merely for acting out the instructions of his genetic code, and now I found myself pulled into his mind, feeling his pain, feeling his fear.

Furthermore, I had saved Darwin only for the time being. I still faced the dilemma of protecting him, and I had few choices, because the use of force, particularly for extermination, had suddenly been taken away. I soon, however, hit upon an alternative solution. I would simply remove the offending stray. Of course, this still left the neighbors' pets, which had diplomatic immunity and could challenge Darwin at will, but I didn't cast my thoughts beyond the immediate problem. Denial is essential if you want to keep your plans on track.

Removal was a practical and practicable plan. Having spent my youth as an avid trapper as well as hunter, I was proficient in the ways of capturing small animals like foxes, skunks, opossums, and cats. The animal pound offered wire box traps for this very purpose, and the next day I rented one. I knew that taking a stray to the pound was essentially a death sentence, but there was always a chance that someone would adopt an abandoned cat, such a big, handsome one at that, and this corrupted rationale was more than sufficient to feed my denials.

That night I placed the trap in one of the paths I had seen the cat take to pass around the garage. For bait I used half a can of fresh tuna, as much to assuage my guilt as to entice the cat, and the next morning I trotted down the stairs to view the trap's yield. It was empty. I was stunned. A stray cat refusing tuna? On the second night I opened another can, this time using premium, unsalted albacore tuna, and used the entire contents. I found my own mouth watering. The next morning I cantered down the stairs, thrust my head around the corner to view the cat close up for the first time . . . and found nothing. The tuna was crawling with ants, which apparently found the bait more appealing than the cat did. That night I used yet another can of choice, white-meat albacore, the most expensive brand I could find.

The third morning I galloped down the stairs, stopped at the bottom, drew a deep breath, and looked around the corner — voilà! — this time the trap had worked. The big tabby stood forlornly in the cramped confines of the wire box and looked at me with an expression I could not read.

As I boy I had trapped some stray cats, and of all the animals I ever caught, including wild foxes, cats were the most formidable. They would crouch down when you approached. The closer you came, the flatter their ears would press against their head, and if you drew within about two feet, they would suddenly launch themselves at your face. The speed was blinding. The strength of claw ripping against wire was difficult to comprehend in a creature this small. Sweat would break out on your face because you knew, in the depths of the primal mind, that if nothing had stood between you and these caged beasts, you would have suffered serious injury.

This cat, however, did not crouch down. His body did not speak aggression or menace. Instead, he pressed his side against the wire and rubbed slowly against it. My suspicions were not allayed; I was not about to be sucker-punched by a quick paw thrust through the wire squares of the trap, and I leaned forward with naked nerves, but at a safe distance, to peer at my prisoner.

I could not help comparing him with Darwin. He conveyed

a wholly different aura, in part because his basic color scheme was dark, and his markings were far more complicated and intricate. His chest and stomach were pastel orange, similar in hue and tone to the fur on Darwin's side and back. A white star marked the center of his chest. Two rings circled each foreleg, and a wider ring formed a necklace just above his chest. His muzzle was dark tan with a network of fine lines on his cheeks, the background color merging into white on his chin and throat. All in all, he was a rakish fellow, his only blemish an inflamed, bloodshot right eye, possibly the result of a fight.

I could not, however, delay the inevitable trip to the pound. But . . . was I absolutely certain that is what I wanted to do? Something gnawed somewhere down below. Had this big, handsome cat hooked a claw into my affections? Of course not. I would proceed with plans in the morning. I carried him in the trap around to the storage room beneath the flat, carefully placed a bowl of water inside, and left him there overnight.

I thought about him that night. He seemed reconciled to his plight and waited in absolute silence when I closed the door and left him in isolation. His cage rested directly beneath the kitchen. I could not get the image out of my mind and drifted off to sleep staring at his face.

Next morning I went down to check on him and found that he had not eaten his tuna and had spilled his water. He had obeyed the call of nature on both channels; fortunately, I had taken the precaution of placing the cage on bricks so the wastes had fallen through the holes in the wire floor.

The time had come to act. The decision had been made. Final. No appeals. I could not keep this cat. It was out of the question. Darwin demanded all the care and attention I could muster. I . . . No. To the pound we had to go. I picked up the trap, carried it and the cat to my car, placed a plastic garbage baggie on the floor of the trunk to protect against urinary discharge, closed the trunk, and headed off.

I drove up Redondo Avenue, directly toward our rendezvous with fate, gritting my teeth at what I had to do, when I noticed that my car was veering toward the right. I compensated by turning the wheel to the left, but the more I turned to the left, the more sharply the car veered to the right. Then it rounded the corner, making a right turn on Anaheim Street. I applied the brakes; the car continued despite my efforts to stop. It proceeded for about half a mile and gradually began to slow, slow, slow, and finally came to a halt — directly in front of the Long Beach Animal Hospital.

I got out, opened the trunk, carried the cat to the front desk, and asked to see Dr. Mader. I had no idea what I was doing.

Fate just happened to have canceled an appointment, so Dr. Mader was available. We entered an examination room, and as soon as Dr. Mader closed the door behind him, I began to explain that shutting ourselves in might not be such a good idea since I had no idea how wild or temperamental this cat was. Dr. Mader took one look, bent over, opened the trap, reached in, began petting the cat, and gently pulled him out, releasing him on the floor. Immediately he began rubbing against the doctor's legs and purring like a power tool. I could feel the fear and anxiety in his dithering passes against the doctor's legs, and I realized with a rapidly reddening face that this was a pussycat, not a street-scarred beast.

Dr. Mader placed him on the examination table and pressed a stethoscope against his left side.

"He's purring too loud. I can't hear his heart."

The doctor persisted, however, and was able to ascertain that the cat's heart sounded normal and healthy. He then placed his hand beneath the cat's chin and looked into his face.

"What's this?" he asked, gently grasping with forefinger and thumb what appeared to be a long eyelash protruding from the outside rim of the right eye. Mader pulled gently, with a slow, gradual pressure, and the eye bulged forward in its socket. Suddenly, in a leapfrogging revelation, I realized that the eyelash was not an eyelash at all: it was the long, slender bristle of a foxtail — a species of grass whose sharp, pointed seeds have the capacity to penetrate the skin and bore into the flesh. The effect can be serious, even deadly, because the seed, driven by the contractions of muscle and flesh against its backward-pointing spines, continues inexorably to bore forward, sometimes entering vital organs. This specimen had somehow moved into the space between the cat's eyeball and socket.

"It has to come out right away," said Dr. Mader, and this could only mean a surgical procedure with sedation. Cash registers tingled like wind chimes in the distance, and I wanted to say no, I could not afford the cost. I said yes without any outward sign of conflict. The surgery would be completed by late afternoon. With time to kill, I got in my car, pointed it toward the pound, and arrived without incident to return an empty trap.

When I picked the cat up at the end of the day, $175 proceeded happily from my savings into the hospital's coffers, and this placed me in a new dilemma. Having spent a large sum of money on the cat's health, I had tacitly acquired another companion, although I refused to acknowledge this fact. I was only doing the humane thing. The cat was still a stray. The most I could do would be to put out some food on special occasions.

Meanwhile, he had to be confined for a few days so I could treat his eye with antibacterial ointment. But where could I keep him? Not in the flat. He and Darwin would certainly fight, and I had no desire to separate two brawling cats in my own living quarters. Then there was the possibility of infection. The FeLV virus was supposed to be highly contagious, and no cat deserved exposure to it. Darwin, however, had always had the run of the flat and had presumably shed virions everywhere.

Well, the risk could not be avoided. The cat had to be closely watched for a few days of convalescence, and as I thought about it, my office came to mind. Darwin had spent little time there, and if I shut the new cat in, he might avoid serious exposure. Darwin would retain 80 percent of the flat for his exclusive domain.

On the way home, I stopped off at the pet store, bought an extra litter box and a matching set of food and water dishes, and set up a temporary household under the south window of my office. When I opened the transport box, the cat set about cautiously investigating the corners and closets of his convalescent quarters. While his attention was diverted, I slipped out, leaving him to his explorations.

Shutting him in would prevent direct conflict with Darwin for a week, perhaps. But what would happen when I freed him? For that is what I had to do. Already I was acting as if this new cat would remain my responsibility, and while I hadn't faced the practical reality, I presumed in a vague sort of way that I would feed the newcomer and care for him outdoors. Only Darwin would have indoor privileges. Even I could see, however, that this was a temporary fix, and Darwin made the point clear that night, on coming in from his daily rounds.

I assumed he wouldn't know the new cat was locked

behind my office door unless he made some fuss. Darwin walked into the living room, stopped, raised his head, sniffed the air, and proceeded directly to the office. He lowered his head, pushed his nose into the space beneath the door, and suddenly the sound of a siren started low and rose to the ceiling, where it swelled in volume and billowed into the living room like smoke.

I hadn't realized how astutely cats perceive odors. Like most people, I had gone along with the public illusion of the dog as *the* master of olfaction, blinkered by the media image of tracking, rescue work, sniffing for drugs and bombs. This was reinforced by the canister prominence of the dog's nose and muzzle, and by its blatant habit of smelling anything and anyone anywhere at the slightest hint of novelty. Somewhere along the way, the cat had lost its own fine sense of smell in the black-and-white simplicity of common perception. Darwin proceeded to disabuse me of this oversight.

Looking back, I began to see the astutness of his nose in countless incidents. Once, for instance, I had tried to trick him by opening a can of tuna as quietly and stealthily as possible to see how long it would take him to realize his favorite food had been sitting around, wasting time. I had even turned the stereo up to overwhelm any sounds I might make. No more than two minutes later, there was Darwin at my feet, begging for his share.

In light of these observations I began to wonder how the cat perceived the world and concluded that as a human creature I was not, nor could I ever be, privy to such sensations. Perhaps, when Darwin entered the flat, the shock of his rival's scent was similar in its negative appeal to finding a skinhead in your living room with a boom box, playing heavy metal rock. Whatever the reality, it was clear that both cats knew at all times exactly who was in the flat, and while the stranger's eye healed, my flat became the house of the rising yowl.

We had a few days, this dark tabby and I, to get acquainted. The first morning after incarceration I came to the door of my office bearing a breakfast bowl of delicious cat food, and before I could turn the doorknob, he became hysterical with anticipation. When I entered the office, he walked quickly in tight circles, rubbing against my legs. He rose up on his hind legs, placed his paws on the dish, and nearly jumped into it before I could place it on the floor. Then he proceeded to bolt his meal with the loud smacking of lips and the slurping sounds of chewing accompanied by suction. He'd had a night to adjust to his new surroundings and his appetite was up. I didn't think any more about that until later in the afternoon when I realized a large rubber band was missing from my desk. Could the cat have *eaten* it? It was then that I began to suspect the presence of a gifted appetite, perhaps a prodigy. I called Dr. Mader, who advised me to watch the cat's stools and make sure the rubber band had passed, because it could lodge in the gut, and that would require surgery . . .

It became clear the next morning when I tended to his eye that the new cat was an affectionate, easygoing creature well rehearsed in stroke appreciation. He made an easy and enjoyable chore of the procedure, resting his head in the palm of my hand to absorb my trepidation and barely blinking as I applied the antibacterial ointment. He accompanied this with one endless purr, like the pedal bass of a church organ — organ music in the truest biological sense.

I kept him in my office for three days and nights, until the eye was beyond infection and the rubber band had passed, and on the fourth morning, after rubbing the antibacterial unguent on his eye, I kneeled down and stroked the sensual lines and elegant patterns of his body and fur, savoring his presence with my hands. So healthy, vigorous, sleek . . .

I could not help comparing him with Darwin and speculating on what might have been. If only I had connected with Darwin a few months earlier, I could have had him vaccinated against the FeLV . . . If only . . . I felt a twinge of resentment — resentment of the burden Darwin had become in his time of need. I spun away and forced myself to face the challenge of keeping the two cats separated.

Feline infectious anemia resolved the issue. On the morning of the fourth day it reemerged and resumed its siege against Darwin's life. When I awoke, I saw Darwin lying on his side at the foot of the bed, gazing vacantly at nothing. His eyes were listless, as if his soul had wandered off into space in search of a somewhere haven, and he would not answer my greetings. I peered into his eyes, but try as I would, I could not push my will past the retina and slip down the optic nerve into his brain. I could not experience his mind. I saw only the vacant, silver-green reflections.

Once more to the hospital. Once more oxygen, intravenous fluids, antibiotic drip, catheter, daily visits, much money. The milk of human kindness. Preoccupied with Darwin's predicament, I freed the new cat. The neighborhood was his for the taking.

Three days later I brought Darwin home for another convalescence, with doxycyclene squirted into the corner of his mouth four times a day for two weeks and pureed food injected down his throat. Somehow, he managed to keep his weight and slowly, gradually, he came back from the brink.

This time, however, the episode left him physically deformed, for his right pupil was dilated more than the left. In darkness or light, the difference remained. Horner's syndrome, said Dr. Mader, common in cats with FeLV. A minor thing, perhaps, but it ruined his symmetry and framed the fear of impending loss.

Another casualty was the posture of his left ear; he now held it lower and pointed more to the front than the right ear. Probably owing to nerve damage, a classic problem attributed to the virus. Both pupil and ear were permanent changes, symbols of time, the passage of life, the solemn gestures of fate.

I let him go outdoors to pursue his interests as nature saw fit. There was nothing to do about the intruder, other than to hope the two did not fight again. As it turned out, they did not, probably because Darwin, in his weakened condition, was spending the majority of his time indoors and simply did not encounter his rival. When he did venture forth it was generally for a few hours during the day, when cats seem less inspired to fight.

Meanwhile, I continued to feed the new cat along the shady north side of the building, and he quickly accepted the arrangement as a natural right, driven in part by an appetite that was turning out to verge on the supernatural. The second time I fed him, he was sitting near the sidewalk at the head of the driveway, and the instant I whistled, he spun and trotted toward me, emitting a long, yodeling cry that bounced and warbled with the jerky motion of his gait — mee-ee-ee-ee-ee-ee-ee-ee-ee-ow-ow-ow-ow-ow-ow-ow-ow — as he trotted along. No matter how often I fed him he wanted more, and I could not help wondering why it had taken him three days to enter the trap for tuna.

The feeding became a ritual that I looked forward to, partly because I felt a kinship with anyone who had such a gargantuan appetite, and partly because I simply felt a deep attraction. With his small, dainty feet, he struck me as a bit of a softy, not so tough in the strategies of the streets and alleys as many of his peers. And perhaps I sensed the random privilege of my own incarnation as a member of *Homo sapiens* and realized in the subconscious labyrinths of my brain that there, but for the grace of God, went I.

Mostly, however, I looked forward to our assignations because it relieved me of Darwin's suffering. For a few minutes I could abandon my worries and cares in the new cat's vigor and escape the aching weight of Darwin's misery. Always this misery waited at the door to my flat like a coat on a hook, required dress for admission to my home, and always I put it on as I entered.

One day several weeks later, the new cat developed a slight limp. My heart lurched and stumbled against the inner walls of my chest, and I reached down and swept him into my arms. He placed his left paw on my right arm and I saw that all his claws were frayed, bloody stubs. I checked the other paw and found the same thing. I had no idea what could have caused this, but whatever it was, it had been violent. I found myself carrying him up the stairs to the flat. I wanted to protect him, to shield him from the dangers of a natural life, but then I stopped midway, remembering that Darwin was sleeping inside. We had come to an impasse. The time had come to face the issue of keeping the new cat.

I sat down on the steps and cradled him on my lap. He commenced to vibrate, the energy passing from his organ of purr into my thighs and proceeding directly into the reptilian complex beneath the cerebrum of my brain. There it caressed the inner lizard of urge, desire, feeling, appetite, lust, and deep, deep pleasure. The lizard understood immediately what the cat was saying, for they spoke directly to each other in the language of touch and feel.

⟳

For years I had been pondering the mind as a process arising from the anatomy of the brain. For years I had been bewildered. I am still bewildered, though not quite so thoroughly, for I had a front-row seat as my mind went into the convulsions of a difficult decision.

As a biologist, I thought in terms of evolution, natural history, plants and animals, physiology, genetics, molecular biology. I knew that certain regions of the brain generate certain aspects of the mind, and I knew that the different regions are interconnected by bundles of nerves which serve as neural pathways.

As a writer I looked for common terms to replace the Brobdingnagian words assigned by science to the brain's anatomy. I wanted names that arose from the nature of the specific regions, based on what they contributed to the mind, and if the name had a wry twist or a certain lilt, all the better. Thus the inner lizard. In time these names assumed a life of their own, like characters in a fable or actors in a morality play.

There was, of course, the inner lizard, named for the reptilian complex or limbic system, the center of visceral impulses like lust, appetite, urge, aggression. This creature was an island unto itself, supremely self-centered and wholly oblivious to the outside world. Having no intellect, it had no awareness of self, no comprehension of space or time, no conceptual grasp of death. It expected immediate gratification; it wanted what it wanted and wanted it NOW. The inner lizard, however, was stitched with countless neurons directly to the other creature, the cerebrum and its cortex, that cap of anastomosing neurons which generated intellect. The reptile, therefore, had to manipulate its cerebral partner because intellect stood between its urges and their satisfaction. The intellect was the lizard's link to the world. Reason operates the hands, the hands manipulate the world, so the lizard, to get its way, had no choice but to manipulate the rational mind.

The cerebum creature was harder to name, in part because there is no antecedent in evolution. There has never been an animal in the natural world without skin, limbs, or muscle; there is no creature incapable of any movement whatsoever, existing as a flabby, defenseless, three-pound gob of neurons that transcends itself in a rationalizing, self-aware mind capable of conceiving existence and death, comprehending infinity and eternity, holding moral values and religious beliefs, aesthetic sense and philosophic curiosity — the whole panoply of intellect by which we humans define ourselves and project our gods. Such a creature could exist only in myth or fairy tale.

I thought of the human skull, that bulbous urn of bone with its high forehead and bulging rump. I thought of Humpty Dumpty, with his immense noggin symbolizing a brain larger than reason, and the name of this fabulous creature came to mind, ready to take its rightful place in the natural history and natural politics of my mind.

The noggingod was everything the inner lizard was not. Where the lizard was supremely self-centered, the noggingod was able to rise above itself as disembodied curiosity and observe the lizard as it writhed in the cold blue light of reason. Because all senses are routed through the cerebral cortex, the noggingod comprehended a world beyond its self; one of its main tasks was to measure, weigh, analyze, and make rational sense of it. As a result of its wide horizons, the noggingod understood that other individuals similar to itself existed outside its skull, and these others had feelings and rights, just as did the noggingod. They had to be respected if one wished to coexist and reap the benefits of social living. Despite what the lizard might feel or want, its actions had to be edited. The lizard had to be curbed.

And so the two regions, two creatures, were joined at the hip by neural bonds, but they were born as natural enemies. Yet despite their differences, they had no choice but to interact, for each needed the other. The noggingod needed the passion and self-centered sureness of the reptile to give it direction and to force the rational mind past the infinite confusion of intellectual choice. The lizard needed the conscious mind of the noggingod to plan and connive and carry out its schemes.

The noggingod entered the mind wars by appealing to truth, beauty, and goodness, to the ascetic fulfillment of self-restraint, to the pride of achieving the long-term goal, to the self-respect of resisting temptation and doing the right thing. Speaking a language known only to the neurons, the noggingod tried to explain that discipline and self-restraint bring a pleasure as satisfying in its way as carnal indulgence.

The lizard, on the other hand, entered the fray with corruption as its guide, attempting to infiltrate the rational mind and convert it to the carnal agenda. When it succeeded, lust seemed supremely logical, right-headed, inspired, even to the moral lizard. In the heat of passion, in the grip of the moment, with no thought of consequences, seducing thy neighbor's wife might seem like a pretty good idea; in the fury of divorce proceedings, murder might seem even more attractive. A master of specious logic, the carnal lizard caused the mind to rationalize, twisting, distorting, magnifying, diminishing the truth. Without the burden of thorough, painstaking thought, this superficial smartness shimmers and shines, dazzles and inspires. In the end the lizard succeeds in turning the rational mind to its carnal agenda, for it has turned the rational mind into the nose ring of reason. Smartness does that in focusing on one issue or another. The cost of focus is denial. Without the most rigorous vigilance for this weakness, smartness becomes a sublime form of stupidity. It is a rare, or old, noggingod that learns this lesson, because the noggingod is also the god of rationalization.

The antidote was wisdom — intelligence tempered by experience. Along with wisdom came morality, a code of restraint extracted from ten thousand years of civilized existence. The purpose of morality is to save us from reason.

⁓

So there I sat, the cat purring on my lap, emotions and thoughts tumbling and turning in a kaleidoscope of conflict.

"WANT . . ." said the lizard, absorbed in its own needs and desires.

"I can't leave him out in the wild like this. He's not tough enough," said the noggingod, corrupted for the moment with righteousness. "There has to be a way. If you want something badly enough, you'll find a way . . ."

The cat gazed outward, serene as the Mona Lisa. I gazed inward by staring outward, with the blank and stony stare of the Sphinx. The cat continued to vibrate. I looked down and savored his young, supple form, his intricate striped and spotted markings, his shiny fur, his metallic eyes. I looked to the top of the stairs and my mind proceeded through the door into my flat. Darwin lay there in a state of misery I could not fathom, with no prospect of recovery. *He was going to die.* Another thought came up from below, spoken with the unctuous tones of a New Age guru.

"You should keep this cat," said the voice. "Darwin has only a short time left, and when he's gone, you'll have this cat to help you cope. You owe it to yourself. You need to take care of *you*."

The logic was compelling. The logic was despicable. I agreed without a whimper. I would keep the new cat. . . . Or, actually, come to think of it, I would *rescue* him; I would save him from a short and brutish life on the street, ending with an early death.

But — what if he was someone's pet? The thought had crossed my mind before . . . well too bad. He would have lost an eye, maybe more, to a foxtail boring into his brain. He had bloody stubs where claws should be. Anyone who paid so little attention to his or her pet was not worthy of this cat and would not care if it disappeared, as it had just done. I carried him purring into my office and shut the door. Moral values were fine, said the carnal lizard, leading me by the nose ring of my own reason; you just had to take them with a grain of common sense.

Decision made, I turned my attention to the impractical matter of incarcerating a cat in a four-room flat, at the same time keeping him away from Darwin. The bathroom and the kitchen received too much traffic to serve as a holding cell, and this left only the room that I used for work. My office, then, would be the new cat's castle. I knew he should have more space than one small room, but I could think of no way to allow it. Perhaps I could give him the run of the flat while Darwin was outside? But that would expose him to the feline leukemia virions that probably lay everywhere.

The new cat, however, seemed to accept his sentence without a second thought. Often, when I entered my office, I found him on the windowsill, stretched out flat on his back, forepaws tucked under at the wrist, floating off in the cosmos of his dreams. So relaxed and contented was his sleep that he never even opened his eyes when I began stroking his belly, my own muscles tensed, nerves strung tight, dreading the convulsion of grasping claws and kicking, raking hindfeet I had triggered in Darwin. Apparently his brain was not wired this way, and he snored away in the ecstasy of a tummy rubbie.

Nonetheless, in the days that followed, guilt slowly seeped into the grooves of my brain because Darwin was not happy with the arrangement. I wanted to evade the guilt and feel good about myself, but I refused to accept the obvious solution and find another home for the new cat. A clever alternative then came to me. What if I could integrate the new cat into the household and create a happy family? By making the two cats love each other, I would account for the needs of both and be absolved of my sin.

The problem of viral infection still remained. Veterinary doctrine held that FeLV was so infectious the new cat would almost certainly be exposed to the disease. There was nothing here, however, that a dash of denial and a pinch of rationalization couldn't fix. I reread my veterinary source books, and found inconclusive answers about the dangers of shared quarters. Although not recommended, use of the same space, furniture, and rugs might not be a death sentence. But shared food and water, along with bite wounds, were far more likely to cause infection. I talked to my cat-loving friends. Oh, they said, our cat so-and-so has been infected for years and none of the others have gotten it. Just what I wanted to hear, and soon my fears subsided. I would feed Darwin and the new cat in different places and not let them fight.

If the worst happened and the new cat contracted the disease . . . well, that would be unfortunate. But so it goes. I was doing this for *me*.

This left only the chore of making two strange adult male cats, enemies to the genetic core, into affectionate friends. If we can put a man on the moon . . . et cetera, et cetera. Counseling would do it. I had seen books about such things. In no time we would be comfortably cohabiting.

The new cat soon revealed a personality that seemed to support my rosy visions of harmony, for he proved affectionate to a fault. He wanted to be near me at all times and took over the surface of my writing desk as his bed. I constructed another nest from cheap bathroom towels and placed it at one corner, hoping to preserve space for my own work, but time and again that did not satisfy him. He would stand up, walk to the edge of the desk, reach down with a forepaw to test my lap, and then, when it passed his standards, he would climb down, curl up, and go to sleep on my legs.

At first I found this behavior a little too needy, even cloying. Darwin had always been more independent and self-possessed, and we had arrived at a low-key, comfortable relationship in which affection was sensed rather than physically expressed. In the beginning he had curled up next to me on the couch while I read or watched television, but my movements, however slight and occasional, seemed to irritate him and after several weeks he chose to lie on the floor, back pressed against my ankles. Even my ankles, however, eventually fell from grace, and from that point on, Darwin curled up nearby but not in physical contact. He seemed to understand that he was firmly entrenched in my affections and that I would accept whatever he wished.

⟶

The matter of a name now became an issue. The new cat had become a member of the family and could not go on in lowercase. The human intellect seems to have an innate need to label its intimates. Strangers, by definition, have no names; friends and acquaintances and, above all, family members must have names. I could no longer refer to this new member of my family as "the new cat" any more than a mother and father could refer to their youngest child as "the new kid."

Choosing a name is a momentous task and takes considerable effort. Having named Darwin a year earlier, I recalled that a good name fits the recipient's character, captures his essence, and, above all, feels right.

As a writer I also had literary honor to uphold, and I wanted to choose a name thematically consistent with Darwin the biologist. Once again the great one came to my aid, for like my own Darwin, he had run afoul af a formidable young rival named Alfred Russel Wallace who recognized the mechanism of natural selection first, thus becoming not only a rival but a nemesis as well. What better name for the new cat than that of a nemesis? Wallace he would be.

I bought a new transport box and labeled it "Wallace Jordan." I began talking to him as "Wallace." But some intangible something was off-key. Try as I would, I could not make the name feel right. The cat was just not a Wallace.

One day, after placing a bowl of delicious cat food before him, I looked down and watched the cat devour his rations. Slurp, snort, smack, lick lick lick lick — I was astounded at how quickly the food dematerialized. When the last crumb had disappeared, the cat licked the bowl so clean I could not tell from my standing height whether it had even been used. It was literally spotless. Then he licked the floor around the bowl. Finally, he pushed it aside and licked the floor where the bowl had been.

Later that day I decided to clean the flat. I wheeled out the vacuum cleaner, flicked on the power switch, the motor roared to life, the vibrations passed up my arm, and eureka! the name blurted from my mouth. His name was Hoover.

⁓

The counseling sessions began a few days later. Darwin had slowly gained strength, and the time seemed right to begin negotiations, placing the two cats in the same room. The first session was almost unnaturally civil. It consisted of me sitting on the couch with Hoover curled up next to me and Darwin sleeping a few feet away on my prized leather armchair. Common sense indicated that the more time the two cats spent together the more accustomed each would become to the other, so there we sat, practicing domestic bliss. The second session went as peacefully as the first. Sleep came quickly to the two patients, and I sat between them, beaming peaceful vibes and retaliatory menace like the enforcer Buddha I was.

The third session began with the same quiet harmony as the first two, with Hoover sleeping beside me and Darwin stretched out on the leather recliner in the angle where the back and the seat joined. Around eleven o'clock — the news had just begun — Hoover stood, stretched first his forelegs, then his hindquarters, walked calmly to the edge of the cushion, dropped to the floor, and ambled lackadaisically toward my office as if he hadn't a care in the world. Judging from his posture and his slow, unconcerned walk, I assumed he was heading to the litter box and turned my attention to the state of the world.

His path led directly past the recliner, and what happened next occurred so quickly that it could have been a dream. Just as Hoover reached the midpoint of the chair, he stopped, whirled, reared up on his hind legs, and threw a roundhouse right at the dreaming Darwin. The blow caught him on the cheek and jolted him awake in a posture that was indefensible. Fur flew. Screeches exploded into the air, and Darwin, caught completely off guard, ended up against the back of the chair — sick, weak, bewildered, cowering, helpless.

If I'd had more experience with the combat tactics of cats I would have known they are consummate sucker punchers, and I could have seen this coming. The cat's entire nature is crafted for the ambush, for that is how these predators hunt, concealed, waiting, strung tight. In times of war, the cat instinctively looks for those precious instants when the enemy relaxes its guard, but I had no experience in these affairs and I, too, was caught off guard.

My eyes saw everything, but my mind could not keep up. Incredulity needs an instant or two. Then, before thoughts could form, my body stood up and a Triassic roar erupted from my throat. The image of Darwin cowering — my sweet, helpless Darwin — triggered a rage I could never have foreseen, and before reason could collar urge, I leaped up and heaved the Sunday

Times — five pounds of scintillating politically correct impending mulch — at the terrified Hoover.

He bolted toward the office and disappeared into the room he knew best. My rage, however, now mushroomed into a wild urge to hurt, terrify, pound into the ground this monster who had attacked my Darwin.

I struggled to control it. I managed to wrestle the thought of serious weapons away from the grip of my fury, but could not stanch the flow of anger and ran to the kitchen for a spray bottle. I opened the nozzle to form a water jet and ran back to the office for the kill.

I rushed into my office, but Hoover was nowhere to be seen. There was only one place he could hide, and there he was, cringing on the shoe shelf at the back of my wardrobe closet. He turned to run, but there was nowhere to go. Swearing lava, I squirted him in the face with the spray bottle. And squirted and squirted and squirted. Finally, as water dripped from his face, my moral, compassionate side was able to wrestle the reptile of rage to the ground and produce pity. I stared dumbly, with rising shame and revulsion at my own behavior.

What was I thinking? Hoover was not responsible for his nature. He had no free will and no moral conscience. For me to act as I did was an unacceptable display of cruelty and stupidity. It left him shivering in shock among the soaked shoes. I went to the bathroom, got a large bath towel, and began drying him off, feeling dirty inside, self-pity and disgust whipped to a foul, frothy slime. It was the last time I ever raised a hand against him, or any creature, in full, unedited anger.

12 | SWEET EPIPHANIES

THE OUTBURST left me staring at the wreckage of my illusions in a vacuum of silence, and I realized I had made a serious mistake in taking on another cat. There would be no happy family in my household. There would be no affection between Hoover and Darwin. Maybe, had Darwin been young and healthy, with life before him, a cease-fire and eventually even a peace could have been worked out, but that was fantasy. My first priority was Darwin's care. Hoover would simply have to endure his confinement, however long that might be.

Darwin never forgave me.

Oh, he didn't reject me on moral grounds, for he had no morality. He rejected me on the practical grounds of survival. Whenever he came in for food, the scent of Hoover must have burned his nose, taunting him with the presence of his archenemy. He had grown too weak to fight this younger rival, and the deep logic of instinct declared that my flat had become a dangerous place for a sick old cat to be.

He began to stay outside and would not come in, even for food. Baffled, I would pick him up and carry him inside; one night, when I attempted this breach of sovereignty, he bit me as

I lifted him from the ground. It hurt my feelings more than my flesh, but it pierced my agendas and made me realize the issue was nonnegotiable. Left with no leverage, I placed his food outside the door.

Dwelling next to the earth, Darwin became prey to fleas, which multiplied in his coat as another symptom of feline leukemia viremia, since the victim loses strength and gradually ceases to groom. Consequently, the fleas prospered beneath the fur and there was little I could do, for even though Darwin had little energy to combat fleas, he still had enough to combat me and refused to allow combing.

In desperation I considered flea powders. These, however, contained pesticide, and in Darwin's weakened condition, that seemed a dangerous and foolish solution. This left no alternative but the occasional flea bath, still with pesticidal ingredients but only in light and occasional doses.

Since I had never bathed a cat before, my friend Robyn of the mystical persuasions came over one afternoon to help. We plugged Darwin's ears with cotton, rubbed a thin layer of petroleum jelly around his eyes to protect the membranes from soap burn, lifted him into the bathtub, wetted him down, and while Robyn gently soaped and lathered him with shampoo, I wrestled and struggled against his stiffly splayed legs and arched claws to keep him in the tub. His struggles, though, seemed halfhearted, almost feeble, compared to what they would have been a few months earlier. I would gladly have traded the extra scratches and puncture wounds for a healthier Darwin. With his fur matted to his skin, his legs looked so thin and so very fragile.

⟿

With Darwin refusing to accept my help, guilt dampened any pleasure I might have had in Hoover's presence. Hoover, sitting calmly, innocently on my desk as I wrote, Hoover with his elegant, sophisticated markings, gazing with motionless eyes into my soul, was the incarnation of my betrayal. I had violated a trust, I had made a double commitment I could not honor; I was a bigamist, and payment was now falling due.

I refused to give up the cause, however, and waited and hoped for Darwin to have a change of heart. It seemed inevitable that sooner or later he would want to live inside because fall had arrived, and cold, rainy weather would magnify the stresses of life on the street, outweighing the repugnance of Hoover's presence. I knew that he *wanted* to be indoors somewhere, and if not with me, then with anyone who would take him in. The landlord's soon-to-be-ex wife, now living in the flat downstairs, confided that Darwin was trying to ingratiate himself with her. His efforts fell like rain on a stone, however, for she had larger worries and cares to address on her own family front.

Then one morning in early October I opened the front door and found Darwin huddled before it. He did not look up. He just lay there, a small, orange orphan huddled on the scab-red carpeting, and I knew without any medical tests that feline infectious anemia had returned for a third attempt on his life. Using the criteria I had formulated on my dark walk to the beach, I attempted to enter his mind by attuning myself to his posture, movements, cries, and divining his wants and needs; my decision on medical care would be his decision as well.

How clearly I felt his mind. Darwin wanted to fight the virus. Of course he did, of course he did. He would certainly consent to the relentless inflictions of medicine and hope in order to live.

Once again I administered pills four times daily. But Darwin's health had lost ground since the last episode, and he could no longer keep the antibiotics down. He refused his food. His weight began to drop.

I responded with a holy neurosis of guilt, love, and grief and watched over him with the focused zeal of a two-eyed Argus in my efforts to divine his needs, know his wants. Wherever he lay I built a bed of towels. Whenever he cried I leaped from my desk and went to his side. I fussed and doted and tried to get him to eat by gently stroking his back. His appetite continued to wane, so I went to the market and bought ever more exotic and expensive morsels to stimulate his interest — shrimp, lox, fresh sole, lobster tail. I tried them all, but nothing brought Darwin's appetite back for more than a few mouthfuls. Dr. Mader suggested baby food, particularly ham, and when Darwin refused even that, I had nothing else to offer.

His weight now dropped steadily, going in two weeks from fourteen to twelve pounds. It was devastating to stand by helplessly, and even though I dreaded the implications, I was finally driven to intercede. If Darwin needed to eat, then I would *make* him eat.

Using a blender I frappéed his food, poured it into a large, 35-cc needle-free syringe, and squirted the liquid down Darwin's throat. The feeding sessions proved to be easier than I had feared, in large part because we had arrived in our relationship at a covenant of familiarity and trust: paws, head, body held firmly, gently from behind, jaws wedged open, food injected slowly, gradually, throat allowed its natural contractions, and always a monologue of singsong sounds to soothe and caress.

We settled into the trials of hospice care, and the conflict between hope and reason continued without respite. The rational eye of reality saw precisely what it gazed upon. It went first to the large, round, black pupil of Darwin's right eye and compared it to the pinched slit of his left eye. It traveled then to the left ear with its strange forward set contrasting with the right ear, standing straight and alert the way the ear of the cat was intended to stand. Finally the eye of reason ran along the ragged coat now beginning to sag over the emerging bones of an ever-thinning body.

With recovery clearly impossible, hope responded by retreating and retrenching. If it could not convince the mind of a full recovery, it could at least soften the unavoidable. It could seize on something attainable and use it like a carrot on a stick to keep us trudging forward. A few days of remission. A normal body temperature. Enough appetite to eat a full meal without assistance. Pathetic hopes, but Darwin's life rested squarely upon them, and as his reality became dimmer and smaller, hope became larger and, eventually, became all.

The pace, however, was not unremittingly gray and dreary, for life became more intense and far more valuable. The sweet sorrow of self-denial assumed its proper role at the center of existence, and I focused all my senses on Darwin's physical appearance. This brought me to a level of communication I could never have imagined.

One day as I scooped him into my arms it struck me that for a cat — perhaps for any animal — the deepest expression of trust was tactile. Being touched often signifies aggression and danger, particularly among wild animals or among strange animals and people. To grow beyond this innate wariness signifies much familiarity.

Darwin's trust had nothing to do with reason, of course, because cats cannot reason in the human sense; it had everything to do with the deepest level of his being and could not be contrived. His trust lay in the deep pleasure of feeling my hand run over his fur, of feeling my finger and thumb rub along his jaw and up behind his ears, of abandoning his posture in the cradling of my lap, and in savoring the sensation of having his belly rubbed without the reflexive impulse to seize my hand and bite it. As for me, trust lay in the deep pleasure of simply knowing that the pleasure passed back and forth between two living things, through fur and flesh, with each intimating his love to the other. The proof was in the purr. The purr, the touch, and the trust were one.

Sounds had their own communion, and as time went on I became preternaturally attuned to Darwin's cries in the night, which sometimes horrified me in their similarity to the wails of a baby. I reacted like a baby's mother. Sleeping soundly, I would suddenly be sitting up, fully awake, hair on end, heart pounding, skin oozing sweat — a long, razor-edged meow having sliced cleanly through a dream and left my nerves bleeding.

I wondered about that. Why did Darwin's voice affect me so deeply? Like any good biologist, I dissected my reaction with a sharp question. Was my reaction accurate? Was Darwin's voice expressing his feelings and moods? Or was I projecting my human presumptions?

Right from the beginning I'd had the impression Darwin was usually irritated when he talked to me. To be blunt, he sounded pissed off. His voice resonated with what seemed to be urgency, frustration, impatience, and eventually I concluded that it must be my own peculiar interpretation. Or perhaps cat and man were wired differently, so that what seemed like irritation to me was nothing more than ordinary ennui to the cat. Why should Darwin live in a constant state of irritation when I pandered to his every need? It made no sense.

But in Darwin's mind it did make sense, because he really *was* pissed off. This became clear one afternoon as he slept before my giant old Bozak speakers. I was gazing absent-mindedly at my friend, so haggard and compromised, when he raised his head and stared quietly straight ahead. Seeing no particular significance in this gesture, I continued to watch, and after a few minutes, Darwin glanced over at me, then toward the kitchen. Still blinded by the obvious, I kept watching as he began to fidget, looking back and forth several times between kitchen and me. Finally, provoking no response, he opened his mouth and produced the most caustic, sarcastic yowl I had heard in some time.

At last my light turned on and I grasped this simple fact: Darwin's first line of communication was to gesture and assume a particular posture. He expected me to understand what the gestures meant. Only when I ignored his silent signals did he appeal to the powers of darkness and come forth with an ungodly wail, particularly at night.

But the cry itself . . . what gave it such emotional force? First, each cry had pitch. With a cat the pitch is high, a soprano voice compared to the basso of, say, a Cape buffalo. Each cry also had a tonal value, like notes from a musical instrument. The oboe with its melancholy wail, the bassoon with its bucolic flatulence, the flute with its soaring, piercing spirituality, the cat with its eerie similarity to the cry of the human infant and its soaring, piercing lack of spirituality.

In addition to pitch, the cry had inflection. Darwin's voice

lilted up or down when he talked, often up *and* down, down
and up, in the spirit of an ululation. The cries had volume.
When Darwin turned the volume up he added urgency to my
reaction.

There was more, much more. If Darwin called two or more
times in succession, he added the element of rhythm to the mix.
Tonality, pitch, volume, rhythm — what was this but the rudi-
ments of music? The only element missing was melody. Of
course, much of modern classical composition also has no melody, yet somehow it passes for music.

And how does music affect us but by moving the emotions?
So Darwin was speaking emotions, not concepts, not words, not
schemes. He was using a syntax and grammar of tone, volume,
pitch, and rhythm to express his desires, his needs, his moods,
and he was conversing in a language akin to modern music.
Much of the time he had no choice but to complain because I
wasn't reading his silent expressions of posture, gesture, and
pose. Since my ignorance struck him as rude, it made for a
noisy relationship.

The kinship of cry and music intrigued me, my intellect having
become a sanctuary from sorrow and pain. Intellect, however,
was not enough, so I appealed straight to heaven and decided to
listen to the B Minor Mass by Johann Sebastian Bach. I pulled a
record at random from the three-disk album, placed it on my
old Thorens TD 125 turntable, and by sheer chance set the tone
arm down on *Osanna in excelsis* from the *Sanctus,* not realizing
that in the *Osanna,* Bach had turned to six-part polyphony with
an eight-voice double chorus to celebrate his insignificance before God, creating the most complex polyphonic music of the
entire Mass. I turned the volume to spiritual levels and stood

before a torrent of ecstasy that soared from my speakers and rose to the skies like the aurora borealis.

Had it been my intention I could not have selected a more sublime passage. The significance of *Osanna in excelsis* — "Hosannah in the highest" — was not lost on Johann Sebastian Bach. It was as if he set out to demonstrate for all time the relationship of spoken language to the primal language of sound that lies beneath words. The score provided three words, and Bach played with them for almost three glorious minutes. He drew the vowels out into long, dancing lines of melody, he drew one melody from another and played them side by side. He gave the different melodies to the various voices of the chorus and sent them circling and pirouetting among the gorgeously machined orchestral parts, particularly the piccolo trumpets, which soared above the voices and other instruments like shafts of holy light. My emotions cannot be described. Those three simple words served to announce the topic, the ecstasy of religious epiphany, but the ecstasy itself came directly from the music. Music is the language of feeling and mood.

Darwin lay without moving the entire time. He never seemed to mind the occasional interludes when I left the earth through music. I looked down at his small form while Bach soared in his heavenly bliss, and a sense of kinship spread over me in goose bumps of revelation. The music of the spheres, the music of the cat — same ancestral soul. These things astonished me beyond words.

The music was very loud, and so far as I could tell, Darwin took no pleasure in it. Besides, I was a writer, not a composer, and the matter of word and sound had dimensions beyond music. Clearly, Darwin was speaking in a syntax of tone, pitch, volume, and rhythm, but how much did human speech owe to the primal language spoken by our animal kin? The old language had not been discarded. It had been built over, like an earlier level of some ancient city, and its meaning thrived beneath verbal syntax in the inflections, lilts, rhythms, stresses, and other elements of spoken sound.

Poetry comes as close to the elements of music and animal cries as the spoken word can and still be speech. So essential to poetry are rhythm, tone, and inflection that its attributes have been given formal names like meter, foot, accent, stress, rhyme, assonance, consonance, alliteration, and so on. Poetry is the form of language nearest to music, yet poetry is not music. It omits melody, possibly because melody wields such emotional force that it tends to overwhelm the rational mind and set its foot a-tappin'. This breaks rational focus and relinquishes the mind's attention to the musical muse. Still, in emphasizing rhythm, tone, and inflection, poetry retains the elements of music that bring emotional truth and diverts the energy that would have become music into the realm of concept and thought and meaning.

Consider a poem by William Butler Yeats.

> The cat went here and there
> And the moon spun round like a top,
> And the nearest kin of the moon,
> The creeping cat, looked up.
> Black Minnaloushe stared at the moon,
> For, wander and wail as he would,
> The pure cold light in the sky
> Troubled his animal blood. . . .

Wander and wail . . . Wander and wail as he would . . . Language, distilled to its essence, is the music of thought.

Exhausted, purged, I looked down at Darwin with a new awareness of him and of me, and understood in the deepest sense that communication is communion. The music of the souls flows back and forth with equal grace. From that point on I spoke to Darwin — to all animals — with a new reverence, for if he did not understand English in syntax, grammar, and meaning, he did understand the tone and the inflection of my speech. I tried to caress him with sound, using words in gentle, lilting strokes to express the joy and gratitude from which the words themselves arose. I was, in other words, speaking the old tongue — the language we use with our infants before they acquire speech and still communicate as pure creatures.

The consequence of this communion was an ever-increasing vulnerability to Darwin's feelings and moods. I had vowed to enter his mind on that dark nighttime walk, but now I found that the process of entry worked in both directions. In boring deeper into Darwin's mind, Darwin penetrated deeper into mine, planting his perceived pain and depression in the center of my being.

This vulnerability came to a climax one afternoon as I sat at my desk, and heightened awareness gradually turned to self-pity. A sense of sorrow spread over my mood and turned to remorse for my sins against animals as a hunter, a scientist, even as a boy. This turned to pity for Darwin, blended with a feeling of victimization: Why him? Why me? Which metamorphosed into the bleakest, most barren loneliness.

The crisis hit without warning. It came at me silently from below, and before I could fend it off, my lips drew back in a silent grin, tears welled from my eyes, and a high, keening whine squealed from my throat. I was crying. A middle-aged man, and

I was crying like death in the forest at night. I cringed, hunched over, tried to dampen this primitive call that cut through walls and windows like the moans of love or the shrieks of murder, but the cry continued. I tried to shut my mouth and bite the sound off, but the grief was as thick and tough as a rubber puck and my jaws could not bite through it. The sound issued from my nose, and possibly my ears, and if I had managed to block all openings, it would have squeezed from my pores.

I felt my grip weaken. Not even the embarrassment and chagrin of neighborhood gossip could stanch this force. It was not meant to be resisted. I just let go, let the pain and anguish of Darwin's impending death wash over me.

As always in moments of deepest stress, my rational mind rose from my body to watch my emotions writhe. What a weird, thin, unholy sound, thought the rational mind. What a bizarre sound . . . and it is coming from me. Very interesting.

My nose stopped up, my chest convulsed, and I cried with an abandon I had not felt since I was an infant. But no infant ever cried like this. No infant could comprehend the chest pains of life with its tragic losses. No infant could comprehend the pain of others. None could imagine the terrifying concept of death. This was a cry for all of that; but pain, tragedy, empathy were superficial to this strange, thin keening. This was the cry of natural history, the cry of an animal, the limbic brain writhing in torment; it was the inner lizard pinned to the ground by the noggingod and forced for the first time to face the truth squarely. Its mate, its comrade in pure existence was leaving. The inner creature could not comprehend. The pain was blind, dumb, without meaning. And so it cried out in pure anguish. There was no Bach in this cry, no Yeats, no poetry, no music.

Just the ceaseless call of grief from a thousand eons before, when evolution first made sound from pain.

The episode lasted for several minutes. Then, as suddenly as it started, it stopped. The sun came out. Except for limpness in every limb, it was as if the incident had never happened. Once more I shouldered our world and trudged slowly along with my friend.

Time for another epiphany, which occurred in two parts, the second so there could be no mistaking the first.

One afternoon Darwin was lying on the landing just outside the front door. He no longer had the strength or the will to go downstairs and patrol his kingdom, so he would meow to go outside and merely lie at the head of the stairs, where he could gaze down like the Sphinx at the small section of the parking area framed in the stairway entrance.

I opened the door to make sure all was well. Darwin did not look up to acknowledge me, even when I called his name, and I knew his misery was building. The question always close to the surface came up yet again. How much longer would he want to go on? When will it be time? How will I know?

I closed my eyes and focused the entire apparatus of my intellect, senses, empathy, compassion, sorrow, incipient grief, love on his forlorn being, at the same time keeping a tight grip on my feelings; I had no desire to break down again, especially in the common space between the two flats.

It was a warm afternoon in late fall, and several green bottle flies hovered and swooped in the air above us, passing the time in the antics of courtship or searching the air for the scent of death and foul food. I paid them no heed. Darwin drew all my attention, and I fought to stifle the sadness of watching him sink.

At which point, without the slightest forewarning, Darwin leaped three feet in the air and plucked a fly from midflight between his paws. This was impossible, of course. With my hypersensitized perceptions and focused intellect I had found entry to his mind and felt his misery. Speaking for Darwin, I could safely say there must be some rational explanation for his incongruous behavior. I went back inside to leave him in peace.

An hour later I looked out again and found him lying just as before. I beamed some more sensitivity, empathy, compassion, et cetera, in his direction.

"Oh, Darwin — . My poor little Dar —"

His ears jerked forward, his body tensed, and he peered intently down the stairs. There was nothing there . . . except for a faint tinkling sound, like tiny bells . . . which was growing louder . . . with a hint of manic energy . . . not so much a tinkle as a rattle . . . And the source then appeared in the door frame below us. A small poodle-ish thing.

Clearly lost, it had good reason to be manic. One of those poor fairies bred to indulge the bonbon affections of the overcivilized, it was a creature whose chances of survival on its own were so minuscule they could not be calculated. Rhinestones sparkled on its encrusted collar. Little bells dangled from the rhinestones, jiggling and tinkling. Little red ribbons clung to its tightly curled topknot. Then this specimen, no larger than a small cat, trotted to the base of the stairs, stopped, and turned its big, brown, trusting eyes heavenward, appealing to Darwin.

Darwin stared intently back. Through the surgical slits in his metallic eyes glared twenty million years of feline malevolence. In one quick movement he stood, arched his back, and expanded to about three times his normal size. There he stood like an electric bottle brush, each hair shorting out and spitting sparks.

Selective breeding had not left the poor bonbon with many useful genes, but it had enough left to realize that something was not quite right with Darwin's picture. It started to whimper and cast bewildered looks from side to side.

I could not reconcile the scene before me with the traditional roles of cats and dogs and watched incredulously as the scene played out. Glaring intently at the poodle, Darwin pitched himself over the edge of the landing and bounded down the steps to do battle. Its one good gene still in operation, the poodle screamed like a dying hare and scampered back from whence it came. (As I suspected, it belonged to a friend of the neighbor next door and lived many happy years, consuming hors d'oeuvres.)

Having routed the civilized offshoot of wolves, Darwin trudged wearily up the stairs and flopped on his side. I sat down to face the failings of reason, particularly the notion of empathic reason as the guide to mercy. Darwin had blasted all that. I was not, despite my best efforts, entering his mind and becoming one with him. There was no question that he was sick and slipping steadily downward, but just as clearly, I was projecting my own angst from what I *thought* he felt.

And that exposed the flaw in my resolution to discard my agendas and merge as pure spirit with Darwin's mind. The truth was, I was not merging with anyone's mind but my own. Finally I saw the ageless circularity: the brain receives input from the eyes, ears, nose, mouth, skin; it processes the input, applying its agendas; then it conjures an illusion of reality, projects this illusion back on the world, sees what it has contrived, and takes the illusion for reality itself. This is the legacy of the biological mind, and there is no transcending it.

The best I could do was what I had always done, and that was to make steely-eyed observations. They were the key. I had to assess these perceptions against the most stringent standards of logic and not go beyond what the observations supported. To do so — to exceed the data — was to project my agenda as truth. In other words, I had come full circle, back to the fundamental task of science: to create an accurate illusion of reality without the distortions of the human agenda. My arrogance laid bare, my conceits stripped away, I looked at Darwin once again with humbled eyes. Communion with another soul could not be manipulated at the whims of human will. Communion required that all agendas be laid aside, all schemes left at the door. It required a blank mind and an open soul that responded in this direction or that, whichever way the truth chose to go.

13 | TENDER MERCIES

THERE IS NO POINT in describing the following months in graphic detail. They stay with me as a blur of hospice chores and anguished endurance, with particular incidents going in and out of focus. The endless skirmishes among reason, hope, and emotional communion grew more intense with Darwin's worsening condition, and I came to view the deepening shadows as a choice between the diabolical mercies of humanity and the brutal mercies of nature.

In my rational moments, when I looked down upon the world, impervious to the corrupting force of hope and emotion, I knew that I had transgressed the natural order of things and was committing my friend to a misery that no living thing deserved. Darwin had progressed so far past the point of natural survival that I had to help him eat. In nature he would long ago have been taken by some larger predator, or died of starvation and his corpse been eaten by scavengers. To our civilized sensibilities that might seem a barbaric and unacceptable line of thought. But had a coyote or a puma brought him home to the earth, death would have been as swift and merciful as any dispensed by medicine.

Even starvation would have been quick in comparison with the medical ordeal he suffered now. Why do we humans pull and nudge and drag our creature companions so far beyond the point of natural and dignified survival? Why do sweet, decent, compassionate, loving people refuse to observe the natural order and consequently raise Hell to the surface for their beloved friends? Finally, I faced the truth because I could find no way to avoid it. The truth was, I helped my fading friend because I could not help myself. Nor could I bring myself to put him down. I just could not bear to do it.

Oh, I still tried to commune with Darwin's feelings, groping to understand his needs and pain, but I no longer had faith in my intuitions. Since I could not trust my feelings and would not follow my reason, I became vulnerable to the doctors and the Hippocratic spirit to which they are sworn. Each time I brought Darwin to the hospital, expecting the doctor to say "It is time," he would say instead, "Well . . . there *is* something else we could try . . ." and I could never say "No," because the treatment might work. We might be able to squeeze a bit more time.

Inevitably, in tiny increments masked by hope and denial, the day arrived when the illusion collapsed and the rational and emotive minds joined as one to accept reality. Darwin was losing much of his food; his abdomen was so tender it could not be touched for examination, probably owing to cancer. His bowels reacted accordingly. Antibiotics no longer had any effect. Nothing had any effect. Christmas was two days away, my birthday three. The days were short, the nights long, darkness and gloom pervaded everything. My humanity then rose to the occasion. If I could not yet do that which must be done, I could at least salvage a small illusion of control and plan the event.

Plans — *realistic* plans — require a grasp of details, for

they aim to manipulate reality. The first step was to cast my mind ahead and imagine the scene. It was a medical procedure, which would take place in the hospital, in the room where so many months before I had watched the two young assistants so gently send that big, handsome cat to his fate. I saw myself holding Darwin while the assistants administered their potions. I saw Darwin grow limp, and I could not bear to look. Something was very wrong with the image. This most intimate moment of our life together, this final farewell — and it was to occur in that bunker of veterinary terror?

But where else? At home, in the flat? Why not have Dr. Mader come to our home and administer mercy there, in our sanctuary? The burst of inspiration, however, flashed back to darkness. The event would have to be scheduled at least several days after the day of decision, and Darwin would be forced to wait and suffer during the interim. More than that, even the doctor's presence struck a dissonant chord. I realized then that during the course of our journey, Darwin and I had made another covenant, an unspoken one, for it was I, and I alone, who should perform the final act. Only I was sanctioned; only I had the sacred right to end our journey.

Now, to modern sensibilities this may seem an unsettling thought. By convention we agree that ending a pet's life is a job for the veterinary doctor, who performs his work with professional skill. I, however, was an experienced hunter and a scientist trained in the methods of animal experimentation. I knew how to kill, I knew how death played out, and I knew I could bring it home with more reverence, love, and intimacy than any doctor.

A friend had mentioned a special pill, a sedative that would knock the pet out. I needed nothing more. Once Darwin was asleep, I would nudge his spirit across the infinite divide with chloroform, applied with the gentlest, dearest love.

Dr. Mader recognized the drug as acepromazine, 25 mg. One pill should be enough, but he would give me two, if, for some unforeseen circumstance, an extra would be needed. He agreed that chloroform should finish the task quickly and humanely. All that remained was the decision, and since I seemed incapable of making it myself, the final call would have to be Darwin's. Finally, a week later, he surrendered to the inevitable.

He had been unable to eat for several days. He could no longer keep water down. For his final stand he had chosen my kitchen table, where I had constructed a thick, padded nest of bath towels, with the edges rolled under to form a ridge around the perimeter. There he had lain day and night, using the ridge as a pillow, staring without expression at some vast, inner vista.

That evening, after the news, I went into the kitchen to comfort him however I could, reassuring him that I would always be there, whatever his needs. But all I could do was gaze with utter despair as he lay on his side, his right paw extended forward, as if reaching out to me. I looked into his eyes and saw two silver-green orbs staring back with an eerie, blank glow. There was no expression, no hint of emotion, feeling, pain. Then, without so much as a facial twitch, Darwin spoke to me.

"It is time," he said.

It was not the childish voice of some animatronic movie creature. The voice was my own — the unmouthed speech of inner monologue that we use in conversation with ourselves.

Darwin had entered my mind on his terms, as an equal, as a friend, and used my own machinery to address me.

I sat down at the table and picked him up, hugged him, held him tight, stroked him gently, and sat for a long time rocking slowly back and forth, my left cheek pressed against his head while a soft, almost imperceptible purr seemed to whisper in the distance.

The next morning I arose as late as I could to prepare the rites. I laid one pill on the kitchen counter and tore off several large pieces of cotton for the chloroform. I dawdled, prepared things several times, tried to stretch time as far as time would stretch. "It is time" ran through my mind as I picked Darwin up, hugged him, set him on the floor. He sat quietly, too weak and miserable to move. I picked up the big red jellybean of a pill with my right hand and tilted his head back with my left. He offered no resistance. I wanted to stop, to put it off just a little longer, but there was no stopping now, no going back. I laid my right middle finger on his lower lip, and he opened his mouth as he had done so many times before. I placed the pill on the back of his tongue and gave it a gentle nudge. It disappeared down his throat and his mouth closed.

Facing directly away from the aching pain of anguish, I reached down, picked him up ever so tenderly, carried him down the stairs in my arms, and laid him at the foot of the staircase so he could lie in the sun while the pill brought sleep. I went back upstairs to busy myself with mindless chores and allow him the final dignity of his own company.

I waited about twenty minutes and readied the chloroform. Then I went down to carry him back upstairs, sleeping peacefully. To my surprise, Darwin appeared fully alert, although he lay in an odd position on his side with his back arched against the bottom step and his head raised to look around.

Perhaps the pill needed more time to work. I walked upstairs again and waited another fifteen minutes. As before, Darwin showed no signs of sleepiness. I waited a half hour more, and this time, as I reached the bottom steps, Darwin began to meow loudly, with a tone in his voice I had never heard but recognized instantly as primal fear. He had moved about ten feet away and was propped against the north wall of the yard in a contorted pretzel of a shape. He tried to move but couldn't. He looked into my eyes and meowed again and again, and finally I realized that the pill was not working as it was intended. Instead of bringing sleep, the pill had somehow blocked his brain from motor control and locked his body in this bizarre posture.

Yet his spirit fought. I could see it in his eyes, in his body as he strained to move. Here he was, contorted in agony and fear, ravaged by a bacterium that destroyed his blood, a virus that destroyed his internal organs, unable to eat or drink, with death from natural collapse a few days off at most . . . and still his body fought to live. It humbled me. I beheld nothing less and nothing more than the force of life. It rose up like a genie from every cell in the body, and it had nothing to do with reason or the power of will. It was more profound than that. I stood before the First Commandment of DNA: Thou Shalt Survive, and for the first time I understood as a visceral feeling beyond intellect, down to the remotest pits of my marrow, that the purpose of life is simply to go on living.

My carefully constructed scenario had collapsed in rubble. So long as Darwin was conscious and mentally alert I could not use chloroform to snuff his life. His trauma would be inconceivable. I rushed upstairs and called Dr. Mader.

"The pill is not working!" I snapped.

"Give him the other pill!" said Dr. Mader.

I stood in a brief trance, holding the phone, while reason and feeling came to terms.

"To hell with that!" I said, struggling to restrain the curses I wanted to scream, and hung up.

I scooped Darwin up in my arms and peered into his eyes. He was not there. What I saw was a creature whose eyes showed not a glint of recognition. The creature wailed in fear and desperation. I rushed around the building, jumped into my car, and sped off to the hospital with Darwin on my lap. Fate was forcing us back to the house of alien odors, cries, pain, and fear, but that no longer mattered. We needed help, and because I had watched the two young assistants send that beautiful cat to eternity, I knew they would do their work with reverence and skill.

The receptionist was expecting us. I reached across the counter and placed Darwin in her hands. I took one last yearning, lingering look at my friend, then the receptionist turned and carried Darwin through the stainless steel doors. And . . . I could not bear to accompany him, an act of weakness for which I will never forgive myself. I had never imagined a debacle like this and had not prepared for it. Even with all the preparation I wished, I would have found the choice agonizing at best. Forced to make a sudden decision, and knowing I would not be able to hide my grief, I took the easy way out and ran away. Besides, I told myself, Darwin was beyond recognizing me or anyone and it would make no difference to him whether I was present or not.

Somehow I managed to sign one release form authorizing euthanasia and another specifying the kind of cremation. By law I was not allowed to bury Darwin at home, which left cremation as the only choice, but I had to choose between mass cremation, in which the bodies of several pets were burned together, or individual cremation, which was, naturally, more expensive but guaranteed that the remains would be only those of Darwin. I could not afford the more expensive, solitary incineration, but I could not accept the thought of Darwin's ashes mixed with the remains of strangers in a common pot. I chose the individual option and would pay for it later.

Then I walked blindly from the office, hiding my face, got into my car, and, with eyelids flapping like windshield wipers, I drove and drove, into a vast emptiness, alone.

14 MISSA FELINA

THE NEXT NIGHT I was invited to dinner by a friend and his wife. They were intelligent, thoughtful people who understood animals, and I knew I'd feel comfortable in their presence, able to withstand the waves of grief that washed over me from time to time. A dinner among friends offered the prospect of relief from my flat, which now throbbed with silence. Hoover's presence did help, but he was not Darwin. No one was Darwin. Late in the afternoon I drove up the highway to the town of Calabassas, letting my eyes wander over the green foothills that turned, as the sun set and the miles passed, into black mountains.

It was a good meal, generously flushed with good wine and marked by the appearance of an opossum and later a raccoon in the kitchen. A good meal that soothed the aching in my chest. On the long drive home, Darwin materialized in the car with me, first on the dashboard, then on the passenger seat, and when I arrived at the flat, there he was, walking out the driveway to greet me. I walked through him — he was only a specter — but as I came to the foot of the stairwell and looked up, my scalp began to crawl, for there, looking down at me, was a pale orange, living cat.

He was a young male, not past his first year, and something in his quick, sensitive movement marked him as feral. Probably born under a porch, behind a trash bin, or in some other urban nest. With the hairs on my neck bristling, I spoke to him the way I would have spoken to Darwin.

"Well, who are *you?* What a handsome boy," and slowly approached the stairs. He looked down at me, warily. I took a slow step. He disappeared down the back stairway.

Generally speaking, I am not inclined to entertain the supernatural, and the creature who had just vanished was clearly of the real world, but his pastel paleness was so ghostlike, and to encounter him at the head of the stairs the night after Darwin's death just as I returned from an outing to avoid his ghost — that was almost too much to dismiss as random chance.

The incident made the real specter all but palpable, for I saw him everywhere. He slept on my leather recliner. He waited before the front door to be let out. He waited in the bathroom while I showered, peering at my dancing image through the shower curtain, met me on the driveway, walking smoothly and intently toward me, looking into my eyes. It was hard to consider him gone.

I desperately missed the feel of his warm, soft, furry body, for touch has memories too. I had always looked with condescension on those who kept lockets of hair or other physical mementos of the dearly departed — how corny could you get? — and I could never comprehend those who stuffed their pets or interred them in little mausoleums. Now I understood. I yearned for some small physical thing that could be grasped and cherished. Then I remembered the glass tube in which I kept the whiskers Darwin had shed around the flat. Words cannot describe the loving touch I laid on that makeshift locket.

I had a week to contemplate life, death, eternity while waiting for Darwin's ashes, and the matter of his apparitions began to interest me. Clearly they were generated by the brain, thus implicating some cerebral mechanism, some internal structure, and probably it was the mechanism of memory.

To process the present instant, the mind needs a record of the recent past. Otherwise it would know only what it saw in the present. Nothing else would exist. To notice change, the mind requires memory. For instance, without a record of the recent past, I could not have perceived that Darwin had been injured, when he returned one night with his right foreleg swollen from a nasty puncture wound. The brain must keep a detailed record to compare against the present, and this implies a complex physiology dedicated to memory, a memory machine.

How would the memory mechanism work? Perhaps by reproducing images as a series of snapshots or a film clip while we were in the presence of the friend, relationship, or whatever. We would not be aware of these images, for attention would be focused on present perceptions, but when the friend, person, or object was no longer there, the images would come forward as memories. In my case, as apparitions of Darwin.

As for the machinery itself, what happens in the brain when a loved one is taken away? Assuming that the brain creates memories by forming or changing molecules, reconnecting nerve terminals, or some combination of these — assuming that something physical happens to record a memory, and assuming further that as time goes on, these changes accumulate, then the process of laying down memories must reflect the creation of structure within the brain.

When a loved one is taken from us, the physiological structure that stores and sorts those memories would no longer have much use. The brain would then dismantle the machinery dedicated to that relationship, altering molecules, severing neurons, connecting them anew, covering old pathways, clearing away the old to make way for the new. Maybe that process causes the pain and sadness we know as grief, for something physical has been amputated without anesthesia.

Thus enlightened, I went back to the hospital at the end of the week to pick up Darwin's ashes. The question of earthly remains had perplexed me for some time. Not knowing it was illegal to bury pets on one's property, I had assumed I would inter Darwin in the small garden plot behind the flat. That, however, had confronted me with a dilemma, because I knew that I would not live forever in this very small unit suited more to religious introspection than to a life in its full, sensual arc, and the thought of leaving Darwin's remains there bothered me. It produced visions of some future tenant preparing the garden, exhuming the bones, and discarding them as mere detritus. Suddenly I respected graveyards, vaults, tombstones, coffins, for human nature wants to preserve memories, and to know that the bones of those we love are safe and protected — to know this sets the mind at rest and tethers the departed souls to the massive crypts and heavy stones where we can always find them.

When I arrived at the hospital, Darwin's file lay on the counter, open to the last page:

E/D
Private cremation
Metal container
Sympathy card

E/D — Euthanasia/Disposal. The final victory of reason over hope. Pain, suffering, joy, love, life itself, reduced to medical statistics. Reduced also to a metal container of ashes, which I held in my hands. Transformed to pure spirit, which I held in my mind.

It rained that night and continued until noon the next day, one of those strong, cold winter storms that make southern California glorious for a few days following, and I used the time to organize Darwin's medical records, from the first to the last appointment. The first was dated 01/10/90; the last, 01/10/91. Darwin died exactly one year to the day of being diagnosed with the feline leukemia virus. Finally, I saw the truth: Darwin had come to me because he was sick and needed help. Right from the beginning there had been little signs — in particular a little cough — but they had seemed so trivial that I dismissed them as harmless.

The wind came up in the early afternoon, and by three o'clock God could not have made a more splendid day. The great, towering cumulonimbus moved overhead in stately procession, like the siege engines of the apocalypse, throwing shadows over the ground and scattering raindrops as they went. Cold and clean, the wind thudded in gusts against the buildings and trees. It stung the skin, numbed the nose. Rays of light pierced the atmosphere on their laser flight from the sun and shattered into the dewdrop diamonds clustered everywhere on the leaves and blades of grass. And Darwin and I walked forth to make things right with the great, blue planet on which we live and die, live and die.

Holding the tin in my hands, I decided to walk around the building and scatter half of the ashes in small portions here and there on Darwin's favorite haunts. That, at least, is what I thought I'd want if I were a cat. The rest I would keep with me as I continued through life. I descended the stairs, and as my foot touched the ground I looked at the canister and saw that I was holding not the ashes but Darwin.

He looked up at me, a strong, healthy cat in the prime of life, the way he was shortly after we met. Fully aware that the mind, in its grief, was playing tricks on me, I decided not to resist the moment, but to let the mind have its way and go with Darwin whither it would. Time to dwell with ghosts. Time to shine the cold blue light into the white heat and fuse reason with feeling, mind with soul, infinity with eternity — to commune with life and earth and Darwin's soul in a dimension where I was a simple living thing, no more and no less than any other.

Darwin gazed quietly up at me, for he was never given to idle chatter. He looked so handsome, so vital, so . . . wonderful . . . and I had abandoned him in his final moments.

"I am so sorry, Darwin. I am so, so sorry. I just couldn't . . ."

"Sorrow is for the self," replied Darwin. Or perhaps it was my inner voice speaking to itself in a kind of split personality. Whatever the truth, it seemed that he was using my mind to speak to me as a species with an intellect equal to my own but with the values of a different world.

"I need no pity," continued the voice. "We all die alone."

He gazed directly into my eyes without any sign of emotion or mouthing of words.

"You saved me from the streets. You gave me shelter, and you gave me time. Time is the measure of life. I have nothing but gratitude."

"Oh, my God . . ." I choked.

"My friend, control yourself," said Darwin. "Tears mean nothing to the world."

"My *friend?*" I said, presuming that nothing had changed between us. "Darwin . . .? It's me —"

"I am not as I appear," replied Darwin.

I didn't want to ask who or what he was, but Darwin replied anyway.

"I am the molecules of memory deep in the brain where spirit and flesh become one. I am you."

He continued to gaze, mesmerizing me with those metallic orange eyes, drawing me through the vertical slits into him.

"Come," he said. "It is time."

I carried him along the north side of the building to the small Japanese garden in the backyard. Ferns grew along the fence, and behind them lay one of Darwin's favorite beds, a shaft of light beaming through the canopy high overhead and lighting the spot like a miniature stage. This would be the first place to leave his substance. I pried the lid from the canister and removed the twist tie from the plastic bag containing the ashes. I peered closely at the stuff and noticed that the ashes were not the black, powdery substance I had imagined. It appeared instead like coarse sand consisting of crushed bone, and, like sand, it was dense and heavy.

I reached in with fingers and thumb to extract a pinch, and the instant I touched the ashes, Darwin spoke.

"Life is Light."

I had no idea what he meant, but since Darwin was emanating from my mind in the first place, he answered my blank questions by reviewing biology — knowledge I had learned but had never applied to the emotions and feelings of my own life.

All animal forms of life have to eat other forms of life, lectured my inner voice at Darwin's request. Only the plant manufactures itself from the unliving substance of water and carbon dioxide. As we break bread, so the plant breaks water into oxygen and hydrogen, and the dioxide of carbon into carbon and oxygen. Breaking molecules is hard work, which the plant performs by taking energy from the sun. From oxygen and hydrogen and carbon, the plant makes the molecule of sugar, and in the bonds among the carbons and oxygens and hydrogens it stores the energy that once was light. Each molecule is therefore a minuscule jar of light. When animal life eats plant life, digestion frees this energy, which drives the chemical reactions known as life — reactions that construct living bodies, bodies that go forth and multiply, eating and fighting and loving and multiplying, but always tracing their existence back to the energy from the sun. Life literally is light.

With new reverence I placed several pinches of ashes on my palm, and as I did, Darwin began to chant:

To the Light, which is Life . . .

On cue I scattered the substance onto the patch of sunlit grass. Then I carried Darwin into the adjacent garden plot, where the moist, brown soil basked in the sun, loosely covered with a scattering of leaves. During the summer Darwin had loved to roll in the dust between the tomato plants, and here too seemed a proper place to consecrate. The storm clouds were still moving out, ponderous and stately before the atmospheric wind, and as I reached in for another portion of ashes, a cloud passed before the sun, the garden grew dark, and a sprinkling of rain began to fall.

"Toss me in the air," said Darwin.

I lowered my hand, preparing to fling the ashes upward, and again Darwin began to chant.

> To the air, which is the breath,
> To the rain, which is the blood,
> I give my substance back.

Truly, the earth may be a living thing. Just as the individual cells of the human brain cannot comprehend the larger mind to which they give rise, so we cannot grasp the notion that we could be but cells in a larger planetary organism, living widgets in a life far vaster than the capacity of a three-pound brain to comprehend. What to us is ecology is physiology to the earth, with water and air transporting nutrients and minerals in rivers and streams, just as blood flows through arteries and veins, transporting nutrients and removing wastes. What to us is evolution is but a single lifetime to the planet, which is born, grows up, grows old, and eventually dies, the lives of us and our kin being transient stops in the endless cycles of energy and substance.

I flung the ashes into the air and watched them fall with the rain back to the earth.

"Mix me with the soil," said Darwin.

I pinched a small pile of ashes from the bag, and with Darwin's mind infusing my own, poured it on the ground.

> And to the earth, the body, the flesh,
> My life returns.

I knelt and kneaded the ashes into the soft, wet earth. As I finished, the cloud, moving fast, passed on and the sun blazed forth in jubilant celebration. I picked up the canister and headed with Darwin to the bougainvillea bush, where we had met, and where now, finally, our journey would end.

The bush was immense, with an arched canopy at least fifteen feet in diameter. The foliage had grown over the past year and now drooped nearly to the earth. I parted it like a curtain and forced my way slowly and very carefully through the tangled mass, showing great respect to the thorns, and stood in a secret chamber of filtered light. The undersurface of the canopy formed a vault eight feet overhead, which glowed translucent green beneath the afternoon sun. Wet leaves covered the ground in a thick, soft carpet, and sunlight, forcing its way through tiny holes in the canopy, cast spots of light that burned on the dark brown leaves. I stood there in this magic grotto, holding Darwin ever so tenderly, and simply existed, without speaking or thinking, for we had entered a chapel of life.

The time had come to say goodbye, and I didn't know how, could not express the ecstasy, the tears of loss and the tears of joy at this communion with existence. I wanted to tell him I loved him but felt the need for something more profound. Darwin said nothing. He lay in my arms and gazed into me.

Soon I felt the answer. It lay in prayer, a skill I had never practiced; my knees were clean.

"Here — put me down," said Darwin, looking at the spot where his bed had been the day we met. With the utmost tenderness I lowered him to the leaves and laid him on his side. He immediately rolled onto his belly, stretched his forepaws out in front, raised his head, and stared straight ahead like the Sphinx, his favorite pose.

"Let the prayer be simple thought," said Darwin.

As if to lead by example, he slowly closed his eyes. Then his head began to nod and gradually drooped toward his paws. I looked down, closed my eyes, and examined my gratitude.

Here I was, a middle-aged man, and I had come to depend for companionship on a cat. The outside world would see my life as a dead end, a transgression of the natural order, a social failure, and strongly suggest that I get a life, meaning a human companion. Mine was also a life to be shunned subconsciously, because to see a fellow human treat an animal with love and respect undercuts the sense of moral decency with which most people seem innately imbued, a decency that, if neglected, erodes one's self-respect.

Civilizations, however, can survive only by exploiting the natural world. In order for economic systems to work, animals and plants — nature itself — must be seen as commodities and never as citizens, because citizenship implies rights. To withhold rights implies wrongs, and wrongs must be denied.

But here I stood, and it seemed like such a natural place to stand. My transgression had been nothing more than treating a cat with respect, nothing more than allowing him into my soul as an equal spirit.

Yes indeed, yes indeed, a grave malfeasance. And what was the harvest of my sin? Why, nothing more than a few insights into my self and a few lessons on life. I decided to count them.

First, Darwin taught me to respect life. He revealed that the essence of respect is self-restraint, and he showed that to respect life you must love life. He taught me in his dying throes that all living things want to go on living, no matter how grave the circumstance, and because I had come to love this small creature, I came to love all living things. As a biologist I understood the need to eat those we love, for life is a cannibalistic affair, with all animal forms eating other forms of life. But as a humbled human being, I now wanted all living things to live as good a life as nature saw fit to allow. Now I saw how brutal, painful, and short life is for those who live honestly, in nature, without medicine, without law, without rights, without any civilized amenity; I would not make it harsher than it naturally is. This love of life consists simply of good will for the living and empathy for their pain, suffused with an immense ecstasy in the beauty and goodness of the earth.

I learned from Darwin that love is like a bank. It solicits our business and encourages us to invest by giving us the pleasures of the body when we are young and vital and vibrating with lust; it pays us back with interest in depthless gratitude for the time that each has given the other over the years, secure in the faith that each will remain for the other until the end goes past, as we face the scheduled pains and oblivion of old age. Thus I learned the lessons of loyalty and personal commitment and self-denial for the benefit of others, for my teacher never nagged and never preached. He drew forth goodness by placing responsibility on me, and me alone.

As for the human mind, I learned from Darwin that it is the universe in a bone, nothing less. And nothing more. I learned from his terminal care that we cannot escape the self, cannot leave the skull, and I learned that consciousness is supremely self-centered, because human consciousness *is* the self. Therefore the grand challenge is to transcend that consciousness and enter the mind of another life, knowing that transcendence cannot be attained.

Ah, but the attempt is the reason for being. Darwin showed

me that the mind is meant to embrace others. It happens naturally, without trying: while the brain makes memories by altering molecules and neurons, the mind grows around those with
whom it lives like roots around a rock. Nothing is more natural
than this process. So I grew around Darwin and Darwin grew
around me, and with this small epiphany the Mass continued.

"Enough," said Darwin. "My work is done."

He curled up, resting his head on his paws, and looked up
at me impatiently from beneath his brow, just as he had on that
fateful day a year and a half earlier, when our lives converged.
Now our time had come to an end.

"Oh, Darwin, I'm going to miss you so much."

"You have much to learn," said Darwin testily. "But you
will. Go with Hoover — you could have done better, but he is
adequate — and with the others you will meet on the way. They
will guide you."

The ineluctable tears, the struggle in the throat, the internal
pressure of exploding grief.

"I will be here so long as you live," said Darwin matter-of-
factly, "here, in the roots of your brain, to visit whenever you
wish. And someday, when your time comes, the molecules of
my memory and your spirit will blend together in the earth. Go
now. I will rest awhile."

He raised his head, stood up, arched his back, and walked
toward me. I dropped to hands and knees and lowered my face
to his. He raised his head and extended his neck in one last gesture of love, and slowly, gently our noses came together. As they
touched, Darwin vanished, and I found myself staring at a small
pile of ashes on the leaves. I stayed there for a few moments, unable to move, feeling the leaves and the damp earth.

I got to my feet, pushed through the wall of foliage, and emerged into the ecstatic beauty of late afternoon. The light slanted into the clouds from the west, exposing caves and crevices and great, soft outcroppings of puffy white. Gulls soared in circles, slicing in white crescents across the towering structures and exulting with wild, cackling yelps in the sheer joy of being gulls.

The air stung my face and I breathed the clear, cold atmosphere, savoring the molecules of oxygen which soon would flow with my blood into my metabolism. Emerging as carbon dioxide, the molecules would rise into the air when I breathed out. I was not standing in the environment; I *was* the environment. I was the substance I breathed and ate and channeled through this earthly form in a promiscuous stream of plasma. I was part of a larger existence that extended forth from the first living things in an unbroken line to Darwin and to me, a live, organic cloth of ever-changing forms that clings to the ever-changing world and will endure till the day the sun goes out. It struck me then that evolution is about kinship and kinship is about love and love is the essence of light.

I raised my face to the sun, drank deep of the sweet living air, and thanked Darwin for giving me Life.

Epilogue

In the end,
because I became a cat,
I became a human being.